RA
976
.P75
2001

WITHDRAWN FROM COLLECTION
Silverman Library – BCC

BERGEN COMMUNITY COLLEGE LIBRARY

D1307826

WITHDRAWN FROM COLLECTION
Silverman Library – BCC

PRIVACY AND HEALTH CARE

BIOMEDICAL
ETHICS
REVIEWS

Edited by

James M. Humber and Robert F. Almeder

BOARD OF EDITORS

William Bechtel
Washington University
St. Louis, Missouri

James Muyskens
University System of Georgia
Atlanta, Georgia

Kurt Hirschhorn
The Mount Sinai Medical Center
New York, New York

James Rachels
University of Alabama
Birmingham, Alabama

Richard Hull
State Univ. of New York, Buffalo
Amherst, New York

George Rainbolt
Georgia State University
Atlanta, Georgia

Thomas H. Murray
President, The Hastings Center,
Garrison, NY

Richard Wasserstrom
University of California
Santa Cruz, California

BIOMEDICAL ETHICS REVIEWS

PRIVACY AND HEALTH CARE

Edited by

James M. Humber

and

Robert F. Almeder

Georgia State University, Atlanta, Georgia

 **Humana Press • Totowa, New Jersey**

Copyright © 2001 by Humana Press Inc.
999 Riverview Drive, Suite 208
Totowa, NJ 07512

For additional copies, pricing for bulk purchases, and/or information about other Humana titles, contact Humana at the above address or at any of the following numbers: Tel.: 973-256-1699; Fax: 973-256-8341; E-mail: humana@ humanapr.com, or visit our Website: http://humanapress.com

All rights in any form whatsoever reserved.

No part of this book may be reproduced, stored in a retrieval system, or transmitted in any form or by any means (electronic, mechanical, photocopying, microfilming, recording, or otherwise) without written permission from the publisher.

All authored papers, comments, opinions, conclusions, or recommendations are those of the author(s) and do not necessarily reflect the views of the publisher.

This publication is printed on acid-free paper. ∞

ANSI Z39.48-1984 (American National Standards Institute) Permanence of Paper for Printed Library Materials.

Cover design by Patricia F. Cleary.
Production Editor: Kim Hoather-Potter.

Photocopy Authorization Policy:

Authorization to photocopy items for internal or personal use, or the internal or personal use of specific clients, is granted by Humana Press Inc., provided that the base fee of US $10.00 per copy, plus US $00.25 per page, is paid directly to the Copyright Clearance Center at 222 Rosewood Drive, Danvers, MA 01923. For those organizations that have been granted a photocopy license from the CCC, a separate system of payment has been arranged and is acceptable to Humana Press Inc. The fee code for users of the Transactional Reporting Service is: [0742-1796/01 $10.00 + $00.25].

ISBN 0-89603-878-5

Printed in the United States of America. 10 9 8 7 6 5 4 3 2 1

The Library of Congress has cataloged this serial title as follows:

Biomedical ethics reviews—1983– Totowa, NJ: Humana Press, c1982–
v.; 25 cm—(Contemporary issues in biomedicine, ethics, and society)
Annual.
Editors: James M. Humber and Robert F. Almeder.
ISSN 0742-1796 = Biomedical ethics reviews.
1. Medical ethics—Periodicals. I. Humber, James M. II. Almeder, Robert F.
III. Series.
[DNLM: Ethics, Medical—periodicals. W1 B615 (P)]
R724.B493   174'.2'05—dc19   84-640015
                        AACR2 MARC-S

Contents

Preface

Western societies generally recognize both a legal and a moral right to privacy. However, at the present time there is no settled opinion in the United States regarding how these rights should relate to medical information. On the one hand, virtually everyone agrees that one's medical records should not be open to just any interested person's inspection. On the other hand, most also agree that some sacrifices in medical privacy are necessary for scientific advancement, public health protection, and other social goals. However, what limits should be set upon those sacrifices, and how those limits should be determined, have long been issues of debate. In recent years this debate has intensified. There are a variety of reasons for this; to mention only three: (1) Over the years the US health care delivery system has become increasingly complex, and with this complexity there has come a need for more and more people to have access to patients' medical records. With each transference of information, breaches in confidentiality become more likely. (2) Medical costs have risen at an alarming rate. This makes health insurance a virtual necessity for adequate medical care, and people worry that they will be denied employment and/or medical coverage if certain sorts of medical information are not kept strictly confidential. (3) Finally, many medical records are now kept in computer files, and the impossibility of guaranteeing confidentiality for files of this sort is a constant worry.

The essays in *Privacy and Health Care* deal with the issue of medical privacy from a variety of perspectives. The articles by Charity Scott and Patrick Boleyn-Fitzgerald are tutorial in nature and introduce the reader to some of the most basic ethical and legal problems relating to privacy and health care. The next four articles exhibit a more pronounced tone of advocacy. The essays by Margo Goldman and by Bill Allen and Ray Moseley both favor greater restrictions being placed upon access to medical information, whereas the chapters by David Korn and Mark Meaney argue for the opposing view. The last chapter in the text, is an Editor's Postscript which summarizes and briefly evaluates the recently promulgated Federal Standards for Privacy of Individually Identifiable Health Information. All of the articles in the text, with the exception of the last, are expanded versions of essays delivered at a conference entitled, "Medical Privacy in the Information Age," that was sponsored by the Jean Beer Blumenfeld Center for Ethics at Georgia State University and held on November 5, 1999.

Privacy and Health Care is the eighteenth annual volume of *Biomedical Ethics Reviews*, a series of texts designed to review and update the literature on issues of central importance in bioethics today. For the convenience of our readers, each article in every issue of *Biomedical Ethics Reviews* is prefaced by a short abstract describing that article's content. Each volume is organized around a central theme. We hope our readers will find the present volume to be both enjoyable and informative, and that they will look forward with anticipation to the publication of *Mental Illness and Public Health Care*.

James M. Humber
Robert F. Almeder

Contributors

Bill Allen • Medical Ethics, Law and the Humanities, University of Florida College of Medicine, Gainesville, Florida

Patrick Boleyn-Fitzgerald • Department of Philosophy, Louisiana State University, Baton Rouge, Louisiana

Margo Goldman • National Coalition for Patient Rights, Wakefield, Massachusetts

James M. Humber • Georgia State University, Atlanta, Georgia

David Korn • Association of American Medical Colleges, Washington, DC

Mark Meaney • Center for Ethics in Health Care, Saint Joseph's Health System, Atlanta, Georgia

Ray Moseley • Medical Ethics, Law and the Humanities, University of Florida College of Medicine, Gainesville, Florida

Charity Scott • College of Law, Georgia State University, Atlanta, Georgia

Abstract

Despite decades of Congressional debates, no comprehensive federal law has ever been enacted to protect the privacy of medical records. Existing state laws represent a widely varying and inconsistent patchwork quilt of privacy protections. Why is it so hard to enact a single, uniform set of laws addressing medical privacy?

The legal debates reflect underlying ethical tensions in protecting the privacy of health information. On one hand, individuals feel strongly entitled to a right to medical privacy. The traditions of medical ethics going back to Hippocrates provide ethical support for such a right. On the other hand, we have long tolerated considerable sacrifices in personal privacy in the name of promoting general societal health and welfare. Examples of the societal goods we have derived from such sacrifices include advances in scientific research, comprehensive public health surveillance and protection, improved law enforcement, and better quality health care.

To date, these societal benefits have justified relatively free access to patient records, reflecting a public policy view that too much individual privacy could be bad for our collective social health. There are sharp disagreements, however, over whether we have made the appropriate ethical and political trade-offs between individual privacy and public welfare.

After Congress defaulted on its deadline in August 1999 to enact comprehensive federal legislation, the Secretary of the Department of Health and Human Services promulgated proposed regulations governing health information privacy. These agency regulations are necessarily limited in several key respects.

Numerous important sticking points that prevented the enactment of federal legislation also remain to be resolved by the agency's proposed rules. This article provides a background for the legal and ethical debates, and explores the areas of sharpest disagreement among privacy advocates, industry proponents, and others with a stake in the information contained in patients' medical records.

Is Too Much Privacy Bad For Your Health?

An Introduction to the Law and Ethics of Medical Privacy

Charity Scott

"I can tell you unequivocally that patient confidentially is not eroding—it can't erode, because it's simply nonexistent."—Mark Hudson, former health insurance company employee.[1]

After years of debate, Congress has still not enacted comprehensive laws to protect the privacy of medical records. Defaulting on its own self-imposed deadline of August 1999 to enact such legislation, Congress tossed this political hot potato to the Department of Health and Human Services (HHS). In her turn in November 1999, HHS Secretary Donna Shalala issued nearly 650 pages of proposed rule-making, the first 550 pages of which is simply preamble to explain her proposed regulations. State laws on medical privacy are so abstruse and intricate—so "extensive, vast, and detailed"[2]—that words commonly used to describe them include "patchwork," "erratic," and "morass."[3]

Why is it so hard to enact simple, straightforward laws to protect the privacy of medical records? Nearly everyone agrees that it is ethically right to ensure the confidentiality of patients' health information. Just how much privacy protection to give that information, however, is a question over which people sharply disagree. Patients often believe that no one except their closest health caregivers should be able to see their medical records without their prior permission. Many others, mostly strangers to the patient, believe they should be allowed to look at those records without first having to ask her permission. These strangers to the patient justify their relatively free access to her medical information on the ground that it is good for her, good for other patients, or good for society at large. Perhaps not coincidentally, such access may also be good for the people accessing the records.

This chapter examines why the ethical and legal debates over medical privacy have proven so intractable for so long. It organizes the competing interests and values in the debates by exploring two questions. First, how much privacy *do* we actually now have? Surprisingly little. This section identifies who currently has relatively free access to our medical records, and discusses many authorized—and unauthorized—uses of the information in those records.

Second, how much privacy *should* we have? This question is addressed from both ethical and legal perspectives. Medical ethics has long championed the privacy of individual patients, but in the name of general societal welfare we have long tolerated considerable sacrifices in personal privacy. In effect, we have assumed that too much privacy could be bad for your—or at least, for our collective—health. This section analyzes the ethical trade-offs that we have made between the benefits of ensuring patient privacy on one hand, and the societal goods we have derived from sacrificing it on the other hand (such as advances in scientific research, protection of the public health, higher efficiency and better quality in the delivery of health care, and even improved law enforcement). This section will also explore how the laws to date have reflected this ethical balancing. A potential third question—how much privacy *can* we have in light of current computer technology and security devices—is reserved for another day.

How Much Privacy Do We Have?

"Privacy has disappeared—get used to it."—Lawrence O. Gostin, health law and policy expert.[4]

"There is a long gravy train forming around our medical records."—Charles Welch, MD, chairman of Massachusetts Medical Society's task force on confidentiality policy.[5]

The Institute of Medicine suggests that an "exhaustive list" of patient record users would parallel the list of everyone associated, directly or indirectly, with the provision of health care, including at a minimum 67 categories of representative individual and institutional users (and multiple individuals within each category may actually have access to the record information).[6] The Congressional Research Service reported that an estimated 400 people may see at least some portion of a patient's medical record during the course of a single hospital stay.[7]

Authorized Access to Patient Health Information

Take the typical hospital record: The information contained in it may be disclosed widely both inside and outside of the hospital.[8] The primary uses of this information are for direct patient care and for billing and payment. In connection with patient care, many health-care professionals (such as physicians, nurses, specialists, allied health personnel, and, in a teaching hospital, residents and students) need the patient's record to provide appropriate medical treatment and nursing care. Many others have access to the record to provide related diagnostic or ancillary services (such as radiology or pathology departments, laboratories, physical and respiratory therapy, dietary services, social services, and discharge planning). In connection with payment, many users both inside the provider's facility (e.g., the billing and accounts offices) and outside the facility (e.g., third-party payers, such as health plans, insurers, managed-care organizations, and even employers and government agencies, as well as the clearinghouses that transmit the claims information) review patient records to determine whether and how much to pay for health-care services under a staggering variety of benefits plans and programs.

Secondary users of patient health information include those who work in numerous supporting services that are important for the efficient and effective functioning of the health-care system (such as quality assurance and risk management, medical education, peer review of individual professionals, accreditation of the facility, and biomedical research). Tertiary users include people or organizations offering private or for-profit services (such as third-party benefits managers, marketers, and database developers). They also include government agencies that collect and use health information to perform their functions, such as public health authorities (e.g., many states mandate the reporting of infectious diseases, child or elder abuse, domestic violence, gunshot wounds, or abortions), bureaus of vital statistics and immunization registries, immigration services, and social welfare services. In addition, schools may require medical information (e.g., information related to allergies, medications, and sports fitness); such information may end up in the courts (e.g., litigation involving malpractice, personal injury, or even divorce, adoption, and child custody matters), and the police and other law enforcement officers may seek patient records in order to investigate crimes against individual persons (e.g., rape, assault, drug use, or drunk driving) and against the government (e.g., under the health-care fraud and abuse laws).

Finally, a lot of medical information is finding its way on to the Internet. Hospitals are experimenting with putting patient records online not only to improve the quality and efficiency of health-care services (for example, to allow clinical information to be quickly accessed in an emergency room across town), but also to empower patients to read their own records and even to challenge the information contained in them.[9] Doctors are going on the Web to communicate with patients through e-mail, as well as to take advantage of Web-based services.[10] Patients themselves are voluntarily putting their health information online—from prescription medications to living wills to electrocardiograms—at such sites as drkoop.com or PersonalMD.com as an aid to their families and doctors in case of emergency as well as to learn more about health topics pertinent to their own situations.[11]

Unfortunately, even with clearly stated privacy policies, health Web sites often fail to protect consumer privacy. A recent report investigating 21 health Web sites has found that there is inconsistency between the sites' actual privacy practices and their stated privacy policies. Moreover, the study found that "(e)ven with the best intentions, many sites do not have adequate security in place to protect consumer information from the casual hacker or someone actively seeking to access company databases."[12] With this introduction, we now turn to the subject of unauthorized users and uses of medical information.

Unauthorized Invasions of Patient Privacy

Although most people are unaware just how routine√°Æand extensively their medical information is disseminated throughout the health-care system, most of the aforementioned uses of medical records are authorized or permitted under current law. It is the occasional well-publicized horror story of abusive or unauthorized access to patient information that prompts sporadic outcries to "do something" to protect the privacy of medical records.

People become most concerned that information about themselves will not be kept confidential when such information is the kind that can be used against them—that can harm their economic, social, or emotional well-being.

Disclosure of sensitive medical information—for example, HIV infection, sexually transmitted disease, genetic predispositions, or mental illness—can cause embarrassment, acute distress, or social stigmatization. Patients' fears of losing the respect and goodwill of others is just one dimension to their fear of losing their privacy. Fear of losing their jobs or insurance or otherwise suffering discrimination in the marketplace also creates motivation to ensure that their medical facts are kept secret. Indeed, there have been reports that people were denied or lost their jobs or insurance coverage when information about their genetic risks was disclosed.[13] Fears of job-related or other economic discrimination are not unfounded in light of a 1996 survey which reported that one-third of Fortune 500 employers admitted using their employees' medical records in making employment decisions.[14] In one well-known account, a banker allegedly was able to use computerized medical records to determine which of his customers had cancer, and he called their loans in early.[15]

Some of the anecdotes recount goofs, or accidental disclosures of medical records. For example, while the University of Michigan's health system was trying to resolve problems with its new patient-scheduling system, it put thousands of patient records on the Internet instead of on what it thought was a secure system. Two months passed before a medical student discovered the mistake.[16] When the Harvard Community Health Plan was computerizing the medical records of its 300,000 subscribers, it inadvertently gave many of its employees full access to the detailed psychotherapy notes of patients.[17]

Other stories reflect intentional, even malicious, invasions of privacy. In Massachusetts, a convicted child rapist got a job in a hospital and by using someone else's password, he was able to access almost 1000 computerized patient records, which he used to make obscene phone calls to young girls.[18] In Maryland, clerks obtained patient information from the state's Medicaid database and sold the data to four HMOs.[19] In Florida, a public health worker took a computer disk with 4000 names of HIV-positive patients and sent it to two newspapers.

Celebrities and political candidates are at high risk for widespread publication of their health information. A hospital worker leaked to a newspaper the fact that the late tennis champion Arthur Ashe was infected with HIV.[21] Although country singer Tammy Wynette had tried to protect her privacy by entering a hospital under a pseudonym, a hospital employee apparently accessed her records and revealed them to the *National Enquirer*.[22] During Nydia Velazquez's campaign for election to the House of Representatives, she learned that information about her previous suicide attempt and subsequent mental health care had been anonymously faxed to the New York media and given widespread publicity. Although ultimately elected, she found the disclosures acutely embarrassing professionally and personally—she had not even told her parents.[23]

One of the most pervasive abuses of medical-records privacy is also often viewed as one of the most innocent: the perennial "browsing" by a health-care organization's employees of the medical records in the system's computers. An employee's ability to access the records easily through a password is often simply assumed to provide the permission to do so. Health-care workers may view patient records "for such diverse reasons as curiosity (e.g., about friends, neighbors, relatives, or celebrities), perversity (e.g., sexual interests), anger (e.g., on the part of an employee who is about to be or has recently been dismissed), or a desire for financial or political gain."[24] Although common, such browsing by insiders is usually an abuse of medical-records access privileges. Sometimes employees may be disciplined for such unauthorized access, but more often their viewing of patient records is not detected nor even viewed as wrong.[25]

The Example of Managed-Care Oversight: Legitimate Disclosures or Unethical Invasions of Privacy?

Distinguishing an abuse of medical privacy from a legitimate need to see a patient's records is often a nice question. What patients and their doctors may view as an intolerable snooping by health-plan representatives may be defended by those health plans as necessary oversight to ensure both high-quality and cost-effective health care. Mental health records provide a good example.

Managed-care organizations (MCOs) often want extensive information about a patient-subscriber's condition before they will pay for extensive therapy sessions or expensive medications. Representatives of MCOs, or the prescription drug benefit companies that have been hired by the MCOs, routinely contact psychiatrists, for example, to review their patients' records. To get authorization for therapy or for medications, psychiatrists complain they have to give highly detailed information about the patient's problems, her symptoms, estimates of how long therapy is expected to last, justifications for prescribing one drug over a cheaper alternative, and so on. According to a Baltimore psychiatrist, "The more specific you are, the more dirty laundry you give them, the more approvals you get."[26] A clinical social worker flatly stated: "These days, you can't just put down the diagnosis—say, 'depression.' No, you have to get very specific, to state that this woman is depressed because she's involved in a masochistic, self-destructive affair with her boss, or that this man has gotten depressed around a sexual dysfunction and hasn't been able to respond to his partner for the past few years."[27]

The MCOs defend the practice as necessary to ensure quality of care at a reasonable cost for their subscribers' mental-health treatment. The observations of the CEO of a large California mental-health management organization are typical. He explains that MCOs are responding to prior histories of "people getting into outpatient therapy and literally spending years there, with no treatment goals and no concern about the resources being expended for the treatment," and he argues that "the increased demand for detailed personal information is part and parcel of what first brought managed-care organizations into being—the need for a sense of discipline, which was totally lacking in the mental health care field."[28] Some physicians also defend the practice of giving MCOs the patient's information as an aid to improving patient care. Said one doctor who switched drugs for several patients after discussing their care and alternatives with a company representative: "I don't see anything ethically wrong with that . . . If the insurance company is paying for medication, that insurance company has a right to know."[29]

Although subscribers have usually signed a blanket authorization form for medical records access when they joined the health plan, psychiatrists acknowledge that patients often are completely unaware of exactly what those forms meant. The tension between loyalty to the unsuspecting patients to keep their most intimate and personal revelations confidential and the psychotherapist's practical desire to get paid for treating them can be quite stark. "I have children to put through college," said one psychiatrist who permitted a records audit without telling his patients. "When I retire, maybe I'll be more brave."[30] Psychiatrists say they now give the equivalent of a Miranda warning to their patients: "Be careful! Anything you say can and will be widely shared and held against you by your insurance company."[31] The American Psychiatric Association estimates that 40% of patients now pay out of their own pockets to avoid these disclosures of their health information.[32]

In light of many people's extensive and easy access to a patient's health information—some with clearly legitimate interests, some with clearly illegitimate ones, and many falling in the large murky middle—we might understandably ask whether "medical privacy" has become an oxymoron.

How Much Privacy Should We Have?

"All that may come to my knowledge in the course of my
profession . . . which ought not to be spread abroad, I will
keep secret and never reveal."—Hippocratic Oath[33]

Americans feel a strong sense of entitlement to health-care
privacy, even though most are unaware how often and to whom
their health information is routinely disclosed. Consumer polls
over the past decade have overwhelming shown that Americans
value privacy highly and believe laws should protect it. According to one 1999 survey, 90% of Americans think that health-
insurance companies sharing medical records with other
companies is an invasion of privacy.[34] Another 1999 poll reported
that 85% of Americans support new federal laws to protect medical privacy.[35] In a 1998 poll, 90% of consumers said it was either
"extremely important" or "very important" to them personally in
choosing a health plan to have confidence that it would keep their
records completely confidential.[36] A 1996 Time/CNN poll reported
that 87% of Americans think laws should be passed to prohibit
health-care organizations from giving out medical information
without first getting the patient's permission. According to an
oft-cited 1993 Louis Harris poll, 85% of respondents said protecting the confidentiality of medical records was "absolutely
essential" or "very important," and 96% thought it was important
that all personal medical information be designated as sensitive,
that penalties be imposed for unauthorized disclosures, and that
laws spell out who had access to medical records and what information could be obtained.[38]

Ethical Underpinnings
of a Right to Medical Privacy

Why do we feel so strongly entitled to a right to medical
privacy? Going all the way back to the Hippocratic Oath quoted

earlier, there are certainly long-standing ethical foundations for such a right. As a philosophical matter, respect for privacy reflects respect for autonomy and for an individual's desires to develop her own sense of self and to shape the relationships she has with others.[39] (Just ask any teenager how important privacy is for the development of identity and personal growth.) As a practical matter, respect for privacy may be justified in light of the harms that can result from disclosures of personal information. A court has recently characterized unauthorized disclosures of patient information as invading two distinct interests: "(1) the patient's interest in the security of the confidential relationship and his corresponding expectation of secrecy; and (2) the patient's specific interest in avoiding whatever injuries will result from circulation of the information."[40]

Depending on to whom the disclosures are made (e.g., to a spouse, a friend, an employer), the harms that result from invading this second patient interest could include severe emotional, social, or economic injury—embarrassment, humiliation, social stigma or marital discord, loss of reputation, and even loss of a job or insurance. If these potential harms are perceived as sufficiently great, the patient may decide to withhold future health-related information from her doctor in the future, thus suffering harm to the first patient interest as well, i.e., impairment in the doctor-patient relationship itself. Indeed, the harms from actual or feared disclosures can be so devastating that many patients take extraordinary steps in their dealings with health-care professionals to ensure that their health information is kept secret. In a recent study, one in six Americans reportedly engages in such "privacy-protecting behaviors" as: switching doctors or paying for health care out-of-pocket; asking a doctor to not write down certain information in their record or to record a less serious or embarrassing health condition; giving inaccurate information in their medical history; or even not seeking medical care in the first place for a health problem.[41] These "privacy-protecting behaviors" or "defensive measures"[42] raise obvious health risks, threaten the integrity of the doctor–patient relationship, and could skew medical research based on patient records.

That many Americans feel their medical privacy is jeopardized reflects their ethical intuitions that they are entitled to have their medical information kept private in the first place. These intuitions are in turn supported by the professional ethics codes that govern health-care professions and professional associations. The professional codes of nearly every health-care profession (e.g., the ethics codes for physicians, nurses, dentists and dental hygienists, mental-health professionals, social workers, pharmacists, and chiropractors) and the ethical standards of numerous health-care professional associations (e.g., for hospitals and health-care executives) all explicitly require respect for the principles of privacy and confidentiality.[43] The codes may refer to privacy or confidentiality as a "core value" or a "fundamental tenet," and usually make respect for privacy a central or guiding principle of the health professions.

If there is so much general agreement among patients and their health-care professionals upon the principle that patients are entitled to medical privacy, then what is the problem with simply drafting our laws to reflect that societal consensus? Although privacy may be valued in principle, it is not considered an absolute ethical right. Like all ethical values, it is often balanced against competing ethical values and concerns. The American Medical Association's Code of Medical Ethics is a typical example of this qualification on the privacy right. The AMA Code begins with a clear affirmation of the privacy right and a strong prohibition against disclosures: "The patient has a right to confidentiality. The physician should not reveal confidential communications or information without the consent of the patient . . . ." However, the Code moderates this language with a qualification on the privacy principle: ". . . UNLESS provided for by law or by the need to protect the welfare of the individual or the public interest." It is this exception—allowing disclosures for the benefit of the patient or the public—that has proven so enormous that some would argue it has swallowed the rule.

The benefits of disclosure to the patient and to the public can be great, as discussed in more detail later. Often, the benefits of disclosures are not readily apparent until after prohibitions against disclosures are strictly enforced. One recent example from Maine illustrates these benefits of disclosure, as well as the proposition that it is possible to have "too much" privacy. In January 1999, the Maine legislature enacted a tough new law prohibiting the release of a patient's medical information without her written permission. This simple and direct prohibition was backed up with heavy fines for violation. The impact was swift and dramatic: hospitals refused to give any telephone information to family and friends inquiring about a patient's status; florists said they could not deliver flowers; priests said they could not see patients for last rites; newspapers said they would be hindered in reporting on accident victims.[45] Even doctors could not compare notes on the same patient without getting written permission from the patient, and clinical labs refused to give patients their results over the telephone.[46] The law was repealed within two weeks, and one of its original drafters commented, "What we really did . . . (is) protect patients more than they wanted to be protected."[47] Maine citizens seem to have concluded that too much privacy could be bad for their health.

The Benefits of Privacy vs. the Benefits of Disclosure: Where Should the Ethical Balance Be Struck?

How much privacy is too much privacy? When should the principle of privacy give way to obtain the benefits of disclosure—both in the patient's interest and for the public's welfare? As the Maine experience illustrates, strict privacy protection laws have their costs. Our society has always thought it was worth making trade-offs between protecting patient privacy and promoting both the patient's and the public's health. This section describes some of the social goods—benefits to the community as a whole—that we have traditionally thought it worth sacrificing our privacy for. Within each category of social benefit, this section also raises an example to illustrate how difficult it often is to resolve, ethically and legally, whether a similar trade-off should continue to be made in the future.

Safeguarding the Public's Health

We have tolerated substantial losses of a patient's privacy when the patient posed a serious health risk to others. We have allowed, or even required through public health laws, doctors to disclose to family members or to public agencies when a patient has an infectious disease that poses a high risk of contagion. So, for example, public health laws aimed at monitoring or controlling the spread of communicable diseases address the reporting of such illnesses as sexually transmitted diseases, HIV and/or AIDS infection, hepatitis, and tuberculosis. Public-health monitoring of patterns of violence has also justified requiring physicians (and others) to report evidence of gun-shot wounds, domestic violence, and child abuse.[48] In some cases, psychiatrists have been held to a legal duty to reveal their patient's confidences, without their patient's consent, by warning others that their patient has threatened to kill or seriously injure them.[49] All of these disclosures without the patient's consent—or breaches of confidentiality, if you will—have been thought justified in the name of protecting the public's safety or welfare. Too much patient privacy was thought hazardous to the public's health.

Yet not everyone agrees that public health measures justify sacrificing individual privacy. One current illustration of the ethical tensions between patient privacy and public health involves childhood vaccination registries. In Illinois, for example, registry advocates would like to expand their computerized database to track the vaccination histories of all children.[50] A comprehensive registry makes it easy for any doctor or nurse to check quickly whether a child's vaccinations are up-to-date, which is helpful in a mobile society where families move frequently, and which might be critical in a sudden visit to an emergency room when the parents may not remember exactly what shots were given when to their child. Families in poorer neighborhoods who may receive less regular and consistent medical care than families in more affluent neighborhoods could especially benefit from such a database. Opponents argue that these health benefits are outweighed by the threats to privacy that vaccine registries pose. Registry critics are concerned that such databases would quickly be expanded beyond childhood vaccinations to the creation of medical files on every American, which ultimately could be used by government, insurers, or others to the detriment of individual citizens.[51]

Facilitating Medical Research

We have also tolerated sacrifices in patient privacy in the name of scientific progress. For decades, medical researchers have collected and analyzed a multitude of data culled from the medical records of thousands of patients, often without the patients' knowledge or explicit consent. The researchers have used this information to advance scientific understanding of the causes, treatment, and prevention of diseases. Has the appropriate ethical balance been struck between patient privacy and medical research if scientific researchers may gain access to this information in patients' records without their express prior permission?

Developments in Minnesota present a case in point. Until recently, researchers in Minnesota were not required to obtain patients' express consent before reviewing their medical records in connection with a scientific research project. This relatively unrestricted access for scientific purposes has allowed researchers at, for example, the renowned Mayo Clinic to monitor the outcomes of care and evaluate the effectiveness of new treatments since the early part of this century. This medical-records research has been credited with, among other things, helping the Mayo Clinic achieve its reputation for high quality care with supporting thousands of studies and publications to advance scientific medical knowledge.[52] Ethically speaking, the clear benefits to scientific progress were evidently thought to outweigh the apparently minimal privacy intrusions into patients' medical records.

The Minnesota legislature changed the law effective January 1997, and the new law reflects a shift in the ethical balance. The new law requires medical researchers to notify all patients in writing that their records may be used for research and to obtain patient authorization for access to their records. One author explains that by this law, "any potential social benefits of epidemiological research were discounted in favor of privacy."[53]

This shift in the ethical balance brought cries of protests. Complaints were raised that the new law would require excessive expense and bureaucratic burden to contact all of the hundreds of thousands of patients seen every year and get their written authorization. Moreover, since the vast majority of patients would probably expressly consent in any event, the law would result in needless costs and burdens on the health-care system. Critics also charged that the law would distort scientific studies, because those patients who refused to consent to giving researchers access to their medical records would be excluded from any study, thus potentially resulting in selection bias.[54] Some of these charges seem to have been borne out in practice. One recent study on the empirical effects of the new Minnesota law found that it resulted in low participation rates and increased time to complete medical research. The study concluded that "(e)fforts to protect patient privacy may come into conflict with the ability to produce timely and valid research to safeguard and improve public health."[55]

Did the Minnesota legislature tip the ethical balance too far in favor of patient privacy? Critics have thought so.[56] Some have urged that a more appropriate balance is struck under the federal regulations for the protection of human subjects in research, often referred to as the Common Rule.[57] Although these regulations provide that researchers obtain the express, informed consent of human subjects prior to undertaking research on them, they also provide for an exception. An Institutional Review Board (IRB) may waive this informed-consent requirement if it finds that the research project poses "no more than minimal risk" of harm to the subjects and that the research could "not practicably be carried out" without the waiver of the consent requirement.[58] As a practical matter, these two conditions allow IRBs often to waive the informed-consent requirement in cases involving medical-records research.[59] Privacy advocates worry that when IRBs judge a retrospective medical-records research project to be socially or scientifically desirable, they are too easily persuaded both that it is "not practicable" to obtain consent from a large number of patients and that the harm to them from such research is "minimal" as well, with the consequence that a "right to privacy that is easily and frequently overridden on grounds of social utility is no longer a right."[60] There is also considerable recent evidence that the IRBs' protections for patient confidentiality are weak both in principle and in practice.[61]

So where should the ethical balance be struck between patient privacy and the advancement of medical research? The controversy over medical-records research has ardent advocates on both sides. The new Minnesota law reflects one balance, and the federal Common Rule involving the protection of human research subjects reflects another.

Improving Health Care Quality, Access, and Accountability

The aforementioned two categories entail trade-offs between an individual's privacy interests and the larger community's interests in public health and medical research. Much of the current controversy over the limits of medical privacy entails a tension between privacy protection and efforts to improve quality of care

—arguably for the patient's own benefit, or for the benefit of similarly situated patients.

Managed-care health plans and other health insurers, for example, say that their unrestricted access to their subscribers' medical records—usually under a blanket authorization form signed by the subscriber upon enrollment in the plan—is the key to controlling health-care costs and to improving health-care quality.[62] To support their claims that their access to your medical information is good for your health, they point to a variety of programs that they can offer based on your (and other health-plan subscribers') medical data. With computers, for example, health plans can compile and analyze large amounts of data to generate statistics—or report cards—on how well they are providing health services, so that consumers may judge the plans' quality and make informed choices about their health. By reviewing patients' medical records, a health plan can also send reminders to its subscribers for mammograms or other routine preventive care, resulting in earlier interventions and ultimately better (and cheaper) care.[63]

Based on patient record review, a health plan also may analyze patterns in certain kinds of care and be able to recommend alternative and improved care. For example, if records show repeated emergency hospital visits by a child with asthma or an adult with diabetes, the plan may be able to intervene and arrange more effective care at an earlier stage than at the emergency room.[64] Many health plans have developed so-called "disease management programs" to encourage subscribers with specific illnesses to take advantage of preventive health-care offerings. After reviewing subscribers' medical records, for example, Aetna U.S. Health Care gets in touch with certain subscribers and their doctors, "asking whether patients with diabetes would like a free kit to test their blood-sugar levels at home or whether congestive heart patients would like a home visit from a nurse to help them stick to a salt-free diet and keep tabs on their weight."[65]

Whether such analyses of medical records are justified by improved quality of care or lowered costs, or whether they constitute unwarranted invasions of privacy, is often a judgment call, and people will disagree over whether the call was correctly made. For example, Harvard Pilgrim Health Care in Boston analyzed their subscribers' medical records to determine which ones had made three expensive emergency room visits. The plan learned that many of these subscribers were alcoholics. The next time they presented to an E.R., their primary care doctors were notified and told to have a talk with them. A Harvard Pilgrim official commented, "I don't know how the alcoholics react when their doctor calls and says, 'I heard you were in your third automobile accident.'"[66]

One highly publicized controversy between protecting a patient's privacy and disclosing her medical information ostensibly for her own benefit involves health plans' use of pharmacy benefit managers (PBMs). When a patient goes to fill a prescription at a pharmacy using her health plan's prescription card, for example, her information is entered into a computer that transmits it instantly to the plan's PBM. The computers check to see, among other things, if the drug is approved, if there are equally effective and cheaper alternatives, or if the drug is safe for this particular patient. The PBM may respond with alternative suggestions for the patient while she is waiting at the pharmacy counter.[67]

The health plans argue that this disclosure of prescription information is justified in the name of quality care and cost control. They say it allows them to recommend cheaper and often more effective alternative medications for their subscribers, with resulting enormous cost savings annually on prescription drugs.[68] Pharmacy benefit-management services also provide a system for cross-checking that patients are not taking dangerous drug combinations, especially helpful when a patient may be seeing multiple doctors who are unaware of the others' prescriptions. In some cases, the system can alert patients and their doctors to the risks of taking certain medications too long. In other cases, it can remind patients when they are due for re-fills for their prescriptions, to aid in ensuring that they are complying with long-term drug regimens.[69]

Privacy advocates, on the other hand, argue that these intrusions into patient privacy do not necessarily improve their health, and cite accounts where plans mistakenly made assumptions about their subscribers' health status solely on the basis of their prescription information. For example, one woman whose doctor prescribed her an antidepressant for sleep disorders resulting from menopause was erroneously encouraged to sign up for the health plan's anti-depression program. Privacy advocates, and even some government regulators, are worried that these PBMs are more concerned with the marketing of, or steering patients to, their parent company's own drugs than with the patient's health. Three of the top PBMs are owned by Eli Lilly & Co., SmithKline Beecham, and Merck & Co.[70]

Where should the balance be struck in medical privacy? Health plans argue it would utterly defeat their efforts to improve quality and contain costs to insist that they get their subscribers' prior consent for every use of their medical record (hence their resort to the initial blanket authorization form). Others argue that we are getting too close to the wrong side of the ethical line that distinguishes medicine from marketing, and that invasions of privacy are becoming justified less by medical ethics (whose goal is the patient's best interest) than by business ethics (whose goal is to make a profit).[71]

Assisting Law Enforcement

We have also allowed intrusions into individuals' medical privacy in the name of public safety. For example, police may put out a description of a fleeing injured suspect and ask that hospitals report any patient who may come to the emergency room matching that description. Federal fraud investigators may comb through hundreds or even thousands of patient records at a hospital to detect patterns of billing fraud in the hospital's claims for Medicare or Medicaid reimbursement. These are viewed as legitimate law-enforcement activities. To the extent they are undertaken without the patient's authorization, however, they are invasions of the patient's privacy. Do the social benefits justify such intrusions?

Law enforcement officers argue strenuously that the public's safety and welfare demand that they have relatively unfettered access to patient records. State police officers or district attorneys have cited instances where quick access to medical information proved critical in apprehending a suspect (e.g., the recent case of a convicted arsonist of at least ten churches across three states), or where privacy protections could undermine prompt investigations of child abuse or domestic violence.[72] Federal law enforcers have testified to Congress that the price of privacy is too high if it would hinder their efforts to fight fraud in the health-care industry. The FBI and Department of Justice are opposed to any new regulation to require them to notify patients in advance that the agencies were going to search the patients' medical records for evidence of provider fraud. These federal agencies are convinced that such a requirement could have a "devastating effect" in their battle against health-care fraud and abuse.[73]

Privacy advocates argue that the balance has been tipped too far in favor of law-enforcement activities to the detriment of individual privacy and health. They point out that, under current federal law, the records of which videos you rented last weekend are given more protection from police snooping than are your medical records.[74] In their zealous efforts to prosecute health-care providers for fraud, law enforcers may be insufficiently attentive to the privacy interests of patients. Last year, for example, the names of 274 patients were matched with their individual laboratory tests on billing statements and were erroneously made a part of the public record in court documents filed by federal prosecutors in Kansas.[75]

Mental-health professionals in particular are concerned that their role in healing is being threatened by demands for their cooperation in policing. Trust in the confidential nature of the psychotherapist–patient relationship is of utmost importance to the therapeutic process. If the patient does not believe her confidences will be kept, she is less likely to reveal them. Even if revealed, the psychotherapist may be disinclined to write them down out of concern they may be disclosed during investigations. Both reactions can jeopardize the efficacy of the therapeutic relationship and the integrity of the medical record itself.[76]

Law as a Reflection of the Ethical Balancing

As the previous four illustrations have shown, we have tolerated significant trade-offs in individual privacy for numerous societal benefits in public health, safety, and welfare. But sharp disagreements remain over whether many of these trade-offs have been worth it, and whether similar trade-offs should be tolerated in the future. These ethical tensions—and our difficulties in agreeing on the right ethical trade-offs—are reflected in the state of our laws, both at the state and federal level. The ethical debates have been aired loudly and repeatedly in Congress in recent years, and many attempts have been made to forge consensus over a federal privacy law. To date, however, starkly differing views over where the appropriate ethical balance should be struck among the various competing interests have stymied all efforts to pass such federal legislation. At the state level, each state has enacted its own different set of privacy rules, often resulting in complexity within each state's laws and lack of uniformity across state lines.

State Laws on Medical Privacy

According to a recent survey of state laws by Georgetown University's Health Privacy Project, only three states have a single, comprehensive set of laws on health privacy. All of the others have adopted their laws on a piecemeal basis over time, with the result that the laws "can be found in nearly every nook and cranny of a state's statutes—in obvious and obscure sections of a state's code, buried in regulations, developed in case law, and detailed in licensing rules."[77] Georgia, for example, has about 90 separate statutes addressing some aspect of health-care privacy—and that number does not include the case law developed by court decisions or regulations created by state agencies.

Why is there such complexity in state laws on health privacy? In part, this complexity reflects the numerous different users, and uses, of health information. States tend to regulate according to each different entity that may collect, use, or disseminate health information.[78] So, for example, Georgia has separate laws governing health information in the hands of: physicians, hospitals, schools, HMOs and other health insurers, nursing homes, pharmacies, researchers, and various public agencies (from the state's benefits programs to the state personnel board to others, such as the coroner's office, Department of Motor Vehicles, jails and prisons, or workers' compensation board). Like other states, Georgia also tends to regulate by specific disease conditions. There are separate statutes addressing, for example, AIDS and HIV disease, sexually transmitted diseases, genetic information, mental health illnesses and disabilities (including mental retardation), and alcohol and drug abuse and treatment.[79] Additional complexity is layered on by the nature of the rules themselves, with some laws requiring that confidentiality must be kept (for example, for HIV or AIDS infection or information considered privileged by courts) and with others requiring or permitting disclosure of health information to public agencies or to others (as, for example, in cases of child abuse, elder abuse, nonaccidental injuries, certain sexually transmitted diseases, vital statistics, spinal-cord injuries, and unusual or suspicious deaths).[80]

Also contributing to the piecemeal approach of state laws on health privacy is the fact that usually laws are enacted over time

rather than as a collection all at one time. They are often responsive to a particular issue that attracts enough public attention
for state legislators to be called on to "do something." Citizens
tend to call on government to right the ethical balance when certain perceived wrongs become highly publicized, and a law is then
passed to correct the perceived ethical imbalance. Last year in
Georgia, for example, a new health-privacy statute was enacted
that prohibits health insurers from selling prescription information received from a pharmacy without a patient's written consent.[81] This new law was perhaps responsive to the media uproar
caused the year before by reports that certain retail drug stores
(including CVS Corp.) had been selling patients' prescription
information to a database marketing specialist, who in turn used
that information both to remind patients to get prescription refills
and to market other drugs by mailing "educational material" from
drug manufacturers to patients with particular illnesses.[82]
Georgia's law reflects the state's position on the ethical issues
raised by this conduct. Although CVS defended it as "good medical and good entrepreneurial" practice, evidently Georgia sided
with those who viewed such conduct as a "breach of fundamental
medical ethics."[83]

Current Federal Laws

" My head hurts from dealing with this issue."—Senate aide
complaining after dozens of unsuccessful meetings to nego-
tiate compromise federal legislation.[84]

"This privacy stuff is its own little monster."—Robert Gell-
man, former congressional staff member and consultant on
medical privacy.[85]

In light of the enormous numbers of differing state laws on
health privacy, it is no wonder there have been calls for a single,
comprehensive, and uniform set of federal laws to govern health

privacy. And yet, in light of the multitude of scenarios in which legitimate claims of patient privacy must be balanced against legitimate claims for access to patient records, Congress has been unable to pass such federal legislation. Although Congress has been hearing testimony about problems in health-records confidentiality since 1971,[86] the federal laws addressing health privacy are still quite limited.

The principal federal privacy law, the Privacy Act of 1974, regulates how the federal agencies may collect, use, and disseminate personal information about individuals.[87] The Freedom of Information Act (FOIA) generally allows citizens access to records held by federal agencies, but it contains nine exceptions that permit the agencies to withhold certain information, including personnel and medical files if disclosure "would constitute a clearly unwarranted invasion of personal privacy."[88] Several federal agencies, such as the Department of Health and Human Services and the Centers for Disease Control and Prevention, have used these FOIA exceptions to resist disclosure of health data and patient or research records.[89]

A few federal laws address the confidentiality of specific kinds of health records. These laws include the federal regulations concerning the confidentiality of human subject research records (the Common Rule, discussed earlier); Medicare regulations requiring hospitals to ensure the confidentiality of patient records;[90] and regulations governing the confidentiality of records of patients who are treated at federally funded facilities for alcohol or drug abuse.[91] Also, the Americans with Disabilities Act requires employers to keep employees' medical records confidential.[92] This law has been interpreted by several recent courts to require the confidentiality of all employees' (not just disabled employees') medical records.[93]

The United States Supreme Court has suggested, in principle, that individuals have a right to informational privacy under the federal Constitution.[94] Like most federal legislation, however, a constitutional privacy right would protect only against government disclosures of information, not disclosures of health information by the private sector.[95] Moreover, the federal courts balance this individual right against competing state interests in public health, safety, and welfare, often resulting in a judicial balancing which, like the ethical balancing described earlier, may tolerate a wide range of government disclosures of individuals' health information.[96] In a limited ruling on the confidentiality of psychotherapists' records, the Supreme Court has recognized a testimonial privilege against compelling the disclosure of communications between patients and psychotherapists (including social workers) in court proceedings.[97]

Proposed Federal Legislation

As part of a health-care reform package known as the Health Insurance Portability and Accountability Act of 1996 (HIPAA), Congress gave itself three years (until August 1999) to enact comprehensive legislation to protect the confidentiality of health information.[98] If Congress did not meet this deadline, then by default it authorized the Department of Health and Human Services to promulgate regulations within six months thereafter (February 2000).

Congress missed its deadline, but not for want of trying to come up with an acceptable package of health privacy laws. Every year since 1996, numerous Congressional hearings have been held and many proposed bills have been introduced in Congress. Before the August deadline, five proposed bills were pending in the House of Representatives, and three were pending in the Senate.[99] From February through April of 1999, extensive hearings were held before numerous Senate and House committees on proposed legislation, offering testimony from a wide range of consumer, professional, scientific, and industry groups. Still, the deadline came and went.

Complying with HIPAA's default provisions, HHS issued proposed regulations in November 1999, and has extended the time for public comment until February 2000.[100] After revisions, they will likely become final regulations with the force of law. These agency rules will be a start, but they are necessarily more limited than would be legislation passed by Congress. Even the Secretary of HHS and the President do not consider them to be a satisfactory long-term substitute for comprehensive legislation that could, and preferably would, be enacted by Congress in the future.[101]

Sticking Points in the Proposed Federal Legislation

What's the hang-up? Why is this so hard? About a half dozen policy questions, to follow, have apparently proven intractable in the Congressional debates. Efforts to compromise on these sticking points simply met with little or no success. Not surprisingly, many of the issues, described earlier, that make it difficult to agree upon the right trade-offs in the ethical balancing between individual interests in privacy and social benefits from disclosure are the same issues that were the stumbling blocks to enacting federal laws. These issues reflect the ethical tensions in deciding, for example, the role that express informed patient consent to disclosure should play in medical research, law enforcement, health-plan operations, and other socially beneficial activities. Other sticking points reflect more formal legal wrangling, such as the debates over whether a new federal law should preempt state privacy laws, and whether citizens should be granted a private civil right to sue for violations of their privacy.[102]

Aside from the principles, of course, there is also the question of money. Much of this debate is not just a matter of reconciling conflicts among competing ethical values. These debates are imbued with high financial stakes. How much is privacy worth in actual dollars? President Clinton has estimated that it will cost the industry around \$3.8 billion over the next five years to come into compliance with his Administration's proposed regulations, while industry officials estimate 10 times that amount.[103] Those people who have had relatively unimpeded access to patient information over the years naturally resist any new obstacles placed in their paths. While their objections may be couched in principled terms—their free access to patient information has been good for the patient, good for other patients, or good for society as a whole—it is also true that many of them have profited from such access as well.

Federal Preemption of State Laws

To non-lawyers, the preemption issue may seem a bit arcane, but it has been a real flash point in the debates. The issue is whether federal law should provide only a floor of privacy protection by preempting (i.e. displacing or superseding) all weaker state privacy laws but not preempting any stronger state privacy protections. Or should federal law provide floor and ceiling privacy protection by completely preempting all state privacy laws— whether weaker or stronger than the federal one—to provide a single, uniform, and comprehensive national law on health information privacy?

Industry groups (and Republicans) supported complete preemption. They argued that being subjected to 50 states' extensive and inconsistent laws on privacy is costly and impractical. They argued that patient information flows across state lines because patients often cross state borders to obtain health care—treatments, consultations, follow-ups, and payments often are not transacted in a single state. One comprehensive federal law would provide much-needed uniformity and predictability, according to such groups as American Health Information Management Association,[104] American Public Health Association,[105] and the Health Insurance Association of America.[106] Medical researchers also supported complete preemption to ensure consistent national rules on access to patients' medical records.[107]

Consumer groups (and Democrats) supported floor preemption, but opposed preemption of stronger state privacy laws. States may have tougher privacy protections than a new federal law on, for example, such sensitive health issues as mental health, sexually transmitted diseases, or genetic testing. If these tougher state laws were preempted, then patients could end up with fewer privacy rights than before the federal law was enacted.[108] Privacy advocates argued that federal law should provide a base-line of privacy protections nationwide, but states should be allowed to experiment with stricter privacy protections than the federal government may adopt. This sort of floor preemption is also supported by the American Medical Association,[109] as well as the National Conference of State Legislatures and the National Association of Insurance Commissioners.[110]

The HHS regulations will provide only floor preemption. Under HIPAA, the agency's regulations may not displace stronger state privacy laws, so the states will remain free to retain and enact tougher privacy protections.[111]

Private Right of Action

Whether individual citizens should be able to sue for violation of any new federal privacy law has been an extremely contentious issue. Democrats and consumer groups favored creating a private right of action against anyone who violated the federal law, arguing that strong enforcement provisions were necessary and appropriate. Republicans and industry groups opposed a private right to sue, however, or at least wanted limitations on such a right, such as a cap on damages or application only to "willful" violations.[112] The proposed HHS rules only provide for criminal and civil penalties. Under HIPAA's enabling provisions, HHS was not authorized to permit private lawsuits for violations, although the Secretary of HHS supports enacting a private remedy for violations under federal legislation.[113]

Under HIPAA, the Secretary is authorized to impose civil and criminal penalties for violations. For unintentional violations, penalties up to $25,000 per year in civil fines could be assessed. For knowing violations, penalties up to $50,000 in criminal fines and one year in prison could be imposed. The penalty could be up to $250,000 and 10 years in prison if the violator tried to sell the health data for commercial purposes or personal gain.[114]

Access to Minors' Medical Records

Another contentious issue involved who should have access to the medical records of minors.[115] If, under state law, juveniles are allowed to seek medical treatment on their own without parental consent, should federal law allow parents to see their children's records involving such treatment? For example, many states allow minors to get medical treatment for sexually transmitted diseases, treatment related to contraception and pregnancy (including abortion), and drug or alcohol abuse without getting their parents' consent or perhaps without even their knowledge. Particularly in the abortion context, these state laws relating to minors' medical and privacy rights are highly contentious at the state level. The issue in the federal debates was whether any new federal law should grant parental access to minors' medical records, or whether it should leave state privacy protections intact.

The HHS-proposed regulations follow the latter route, so that minors will continue to have whatever privacy and treatment rights the states choose to give them. If parents have the right under state law to see their children's records, then they retain that right. Under the federal regulations, juveniles who are allowed by state law to consent to treatment on their own will have the same privacy rights as adults, but the states still retain the underlying authority to decide when minors may consent to treatment without parental involvement and whether parents may see those records.[116]

Access by Medical Researchers

The issue over whether medical researchers should have to get the express informed consent of patients prior to undertaking retrospective medical-records research surfaced with a vengeance during the Congressional debates. Medical researchers believe that medical-records archives are a very valuable national resource. Citing the Minnesota example and the arguments discussed earlier in this chapter, they view having to get patient consent as a threat to scientific progress in terms of excessive cost, burdensome bureaucratic red tape, and the potential for invalid research results through selection bias (e.g., by excluding the records of patients who refuse to consent, who cannot be located, or whose express consent cannot be obtained for some other logistical reason).[117] They urged Congress to adopt privacy laws for medical-records research that would reflect IRB-like review, and would allow researchers to proceed without express informed consent, if it was impractical to get consent and confidentiality safeguards were in place for the research project.[118] Health plans and pharmacies also objected to extending IRB protections too broadly by imposing costly confidentiality duties beyond the realm of traditional scientific research projects to reach many routine health-care operations, such as outcomes research, disease management programs, or other activities aimed at improving quality of care.[119]

Privacy advocates supported the idea of expanding the scope of federal laws addressing confidentiality in medical-records research, because currently the IRB regulations apply only to federally funded research. Some have also worried, however, about the apparent weakening of the informed-consent principle in IRB protocols concerning waivers, effectively turning patients into research subjects without their knowledge.[120] Many privacy advocates believe that IRB protections are weak in principle and have proven weaker in practice, and should be made tougher before being extended to private-sector research projects.[121]

Echoing the views of the scientific community in its formal recommendations to HHS, the National Committee on Vital Health Statistics concluded that "requiring patient consent as a condition of researcher access is impractical and expensive. It would also most likely stop a significant amount of useful investigation. This is not in the health interest of individual patients or the general population."[122] The HHS regulations reflect this position. They would allow health-care facilities to disclose health information without patient authorization to researchers whose protocol has been reviewed and approved by an existing IRB or a newly created "privacy board." In effect, they would extend the federal Common Rule to all researchers, whether in federally funded or private-sector research.[123] Researchers would be allowed to access records without patient consent if the research could not "practicably" be carried out if individual consent were required; if the disclosures involve no more than "minimal risk" to the research subjects; if the research is important enough to outweigh the intrusion into privacy; and if the research project has adequate confidentiality safeguards.[124]

Access by Law-Enforcement Officers

Current federal law does not require federal law enforcement officers to obtain a search warrant or even to notify patients before getting access to their medical records, nor does it mandate any other procedural or judicial protections.[125] Consumer groups and physician organizations wanted new federal privacy laws to create tougher standards before law enforcement officers could get access to patients' medical records. Many doctors are uncomfortable with being asked by police to violate their patients' privacy. They would prefer that police be legally required to go through some judicial review process—e.g., get a warrant or other court order—before the doctors are required to release their records to the police.[126] Civil liberties groups also urged that law officers should have to meet tougher federal standards than at present.

The ACLU argued that, as with other areas of individuals' right to be free from unreasonable government intrusions into their privacy, the police should have to show "probable cause" that a crime has been committed before gaining access to patients' records.[127]

Law enforcement officers have argued that federal law should maintain the status quo and impose no new privacy protections, such as requiring a court order or other review prior to gaining access to medical records. They wanted exceptions in the case of health oversight activities, such as fraud and abuse, and they wanted to preserve their ability to get immediate access to medical information without a warrant in emergencies, such as bomb threats or hostage rescues. Law enforcers argued, for example, that they should be allowed to get local hospitals to notify them immediately if a fleeing suspect showed up for treatment of a gunshot wound, or to obtain the health information about someone who is holding hostages or is being held hostage.[128]

Secretary Shalala initially appeared inclined toward the status quo, but the HHS proposed regulations do enact some new protections. HHS has no authority to control local, state, or federal law-enforcement officers, but the regulations allow the health-care facilities to insist upon, in some circumstances, certain minimum legal procedures—e.g., a warrant, grand jury subpoena, or administrative subpoena or summons—before having to disclose confidential information during a law enforcement inquiry.[129] At a minimum, an officer would have to seek an administrative subpoena, which is issued internally by the law-enforcement agency. Privacy advocates hope that this provision will be strengthened in the final regulations to require that approval from a judge or magistrate be obtained before release of records. The ACLU argues that since an individual's privacy in her home is protected under the Fourth Amendment by the requirement of a search warrant before government can intrude, medical records in a doctor's office should be given the same protection of prior judicial review.

Informed Consent or Patient Authorization

Finally, there was substantial controversy over the nature of the consent, or authorization, that health insurers and managed-care companies should be required to obtain from subscriber-patients before using their health information in the course of treatment, payment, and—most controversially—"other health-care operations." Current practice is for health plans to get a signed blanket authorization form from subscribers allowing them pretty much to do anything reasonably foreseeable with their records. Some of the proposed federal bills made the granting of such authorization a condition of enrollment in the plan—if the subscriber would not give such blanket authorization upon enrollment for use of his records in the future, the health plan would not be required to sign him up.[131]

Consumer and physician groups opposed this practice as "coerced consent"—forcing patients to consent to widespread disclosures of their medical information on penalty of losing their health insurance. While patients might be asked to consent broadly to disclosures related to treatment and payment, privacy advocates argued that express, specific consent should be required for "other health-care operations." The industry in turn vociferously complained that it would be impossibly inefficient and unbearably bureaucratic to go back to each patient for each additional use beyond payment and treatment. Such a burdensome requirement, argued industry opponents, would impede their efforts to improve quality of care and control costs through such activities as disease management programs and coordinating care across health systems.[132]

The proposed regulations do not require either patient consent or authorization before disclosures of patient information can be made in connection with treatment, payment, and health-care operations.[133] They simply provide for notification to the patient that such disclosures may be made. So, for example, express patient consent to disclosure of the information in their medical records would not be needed for such health-care operations as quality assurance, utilization review, credentialing, developing clinical guidelines, and auditing for fraud and abuse.[134] Although there is no opportunity for a patient to forbid these disclosures, the regulations at least go further than current practice in alerting patients to what information is going to be disclosed, who would get the information, and when any specific additional authorization for disclosure given by the patient would expire or could be revoked. Also, any further disclosures beyond those listed would require express patient consent, and health-plan enrollment could not be conditioned on such consent.[135] So, for example, a patient's express authorization would have to be obtained before her information was disclosed to a marketing firm or to her employer for use in making employment decisions.[136]

Some physicians and consumer groups strongly oppose these regulations to the extent that patients are merely notified about, without the opportunity to object to, this broad range of disclosures by health plans of their information as it may relate to treatment, payment, and health-care operations. Firmly of the view that the regulations give health plans too much leeway to use and disclose patient records without true patient consent, several doctors' groups stayed away from the White House ceremony that unveiled the proposed regulations.[137]

Other Provisions in the HHS Proposed Rule

Although it is likely that the proposed rule will be finally promulgated, nearly everyone would still prefer that these regulations remain only interim measures until Congress enacts comprehensive legislation. Even the Secretary of HHS acknowledges that the proposed regulations have at least two inherent limitations. First, they do not cover all health records. Under HIPAA, the regulations may cover only electronic records, not paper records. They were written to apply to any health information that is or has been in electronic form at any time (so, for example, if there exists one paper copy and one electronic copy of the same information, the regulations apply to both).[138] Information that is solely recorded in paper form would not be covered by the regulations, however. Most records today are still in paper form, although the vast majority of records are expected to be electronic over the next few years.[139]

A second inherent limitation in the proposed regulations is that they do not cover everyone who holds patient information. They cover health information only in the hands of specified "covered entities": health-care providers, health plans, and health-care clearinghouses (entities that process health information between provider and health plan, e.g., billing). Entities that are not "covered" include a multitude of entities who may receive health information from a covered entity in the course of business dealings with it, such as lawyers, auditors, consultants, data-processing firms, billing firms, and private accreditation organizations.[140] Because HIPAA authorized HHS to make rules only for health plans, clearinghouses, and providers, but not for the many entities who might receive information from them, the regulations cannot regulate directly how these other entities may use or re-disclose information they receive from a covered entity.[141]

HHS proposes to solve this regulatory gap through "business partner" agreements. HHS will expand privacy requirements through its proposal that all the business partners of a covered entity must be contractually bound to adhere to the regulations. Thus, the contract between a covered entity (provider, health plan, or clearinghouse) and a business partner must provide that the business partner may not use health information in ways that the covered entity could not have used it under the regulations. So, for example, a business partner could not sell information to a financial services company because a covered entity could not do so under the proposed rule.[142]

Moreover, the covered entity would be held accountable for any business partner's breach of the contract that it knew about, or should have known about, and did not try to cure.[143] HMOs and other covered entities are not happy with this proposal, since it potentially holds them responsible for someone else's misbehavior (e.g., a drug store with whom they contract).[144] These so-called "chain of trust" agreements, however, do allow HHS to attempt to extend the regulatory privacy protections beyond the covered entities it was directly authorized to regulate. Congressional legislation would not face such restrictions and could cover directly all entities who hold health information, which is one reason HHS continues to prefer passage of federal legislation to its own proposed rule.

Other highlights of the regulations include a number of fair information practices. Patients would be given a federal right of access to see, copy, and correct their records.[145] Moreover, they would have a right to receive an accounting of all instances when their information was disclosed for purposes other than treatment, payment, or health-care operations.[146] Health plans and providers are also restricted to making the "minimum necessary" disclosures—they may not use or disclose the entire file when a smaller part may serve the purpose.[147] For example, an employer requesting information from a provider about a work-related injury might currently be sent the entire file. That would no longer be possible under the regulations.[148] Banks or credit-card companies who handled payment for medical care could not obtain treatment information.[149]

Perhaps most importantly, the proposed regulations would require providers and health plans to maintain administrative and physical safeguards to protect confidentiality and protect against unauthorized access. These safeguards may sound bureaucratic and boring, but they may be key to developing a culture of privacy within a health-care institution. So, for example, covered entities will be required to have a set of written privacy policies; to designate a "privacy official" within the institution to develop policies and procedures concerning use and disclosure of health information; to provide training to employees; to develop administrative, technical, and physical safeguards against unauthorized uses; to develop a complaint mechanism for violations; and to impose disciplinary sanction for violations by both employees and business partners.[150]

Conclusion

"Experience should teach us to be most on our guard to protect liberty when the government's purposes are benefi-cent. Men born to freedom are naturally alert to repel inva-sions of their liberty by evil-minded rulers. The greatest dangers to liberty lurk in insidious encroachment by men of zeal, well-meaning but without understanding."—Justice Louis D. Brandeis[151]

Your privacy is a lot like your reputation—once you lose it, it's very difficult to get it back. Both the ethical tensions and the legal debates have entailed considerable balancing of an individual's interest in privacy against other larger concerns of the community. Have our social ends in the promotion of public health, safety, and welfare too quickly justified the means of sacrificing personal privacy? Some have thought so. Beverly Woodward, a Brandeis University sociologist, has worried that at times, "it has seemed that individual privacy will get balanced right out of existence."[152] During the debates on the proposed legislation, Richard Sobel, a research fellow at Harvard Univer-sity, commented, "I'm afraid the compromise bill will compro-mise privacy. . . . You can't have just a little bit of privacy."[153]

Overall, the argument that too much privacy was bad for our collective health has been a pretty successful one. We have often been persuaded that maintaining privacy was not in the patient's or in the public's best interest. We have often agreed that letting the *patient* control who has access to his health information would be, for example, bad for the health of the individual patient (e.g., by preventing MCOs or PBMs from ensuring that the care she receives is appropriate and cost-effective); bad for the health of other patients (e.g., by threatening scientific advances through medical-records research); bad for the health and safety of other people (e.g., by obstructing public health reporting and surveillance measures); bad for the American public in general (e.g., by impeding law-enforcement efforts); and bad for the viability of the American health-care system as we know it (e.g., by derailing efforts to improve its quality and lower its costs).

Perhaps the ethical and legal trade-offs have been the right ones, as surely many people would argue. I would like to make a few cautionary observations, however. First, as Justice Brandeis warns, we should be wary when someone (especially government) tells us that they are doing something for our benefit, that "it's for our own good" that we are being asked to give up something that was ours, be it our liberty, our property, or our privacy.

Second, we should be wary when someone other than the patient argues that it is too much trouble—too costly, too burdensome, too impractical—to ask the patient what she or he would prefer to do. Medical researchers, HMOs, law officers, public health officials, and others all have strenuously argued that it is just too much trouble to get the patients' voluntary and informed consent before they look at the patients' records. The problem with this argument is that we have heard it before, in contexts that now we look back on and think how wrong it was not to have gotten the patients' consent at the outset. For example, physicians long resisted any duty to obtain the patient's informed consent to treatment, on the grounds either that they (the doctors) knew what was in the patient's best interest, or that requiring consent might impede the treatment itself (the patient might refuse). Yet now there is general societal consensus that, ethically speaking, it is the patient's right, not the doctor's, to decide what treatment the patient should get. The law in every state now reflects that ethical view by requiring that patients be given the opportunity to be informed about, to consent to, or to refuse medical treatment before it is rendered.

Moreover, the history of human subject research in this country should remind us of the price paid by sacrificing the informed-consent principle too quickly. There are numerous examples of scientific experiments where researchers deliberately failed to obtain the prior consent of the human subjects who were studied—such as the infamous Tuskegee Syphilis Study or the Cold War radiation experiments. These activities are now regarded as clear abuses of individual rights rather than celebrated for any scientific achievements or other social goods that may have originally motivated their sponsors.

Finally, rather than being bad for your health, assurances of privacy may be the most important foundational element for promoting good health. As was discussed earlier, studies have shown that patients often engage in "privacy-protecting behaviors"—such as withholding information from their doctors, getting doctors to compile incomplete or inaccurate records, going to several doctors, or avoiding health care altogether. These behaviors can threaten both the doctor–patient relationship and the quality of health care provided. Assuring privacy within the doctor–patient relationship encourages full and open dialogue, which is essential to appropriate diagnosis and treatment. Trust may well be the first premise of high-quality health care.

Along this line, the United States Supreme Court has recognized that privacy laws that promote an individual's health also promote the public interest. In the context of requiring that communications between psychotherapists and patients be kept confidential, no matter what bearing they may have in the search for truth in court proceedings, the Court observed: "The psychotherapist privilege serves the public interest by facilitating the provision of appropriate treatment to individuals suffering the effects of a mental or emotional problem. The mental health of our citizenry, no less than its physical health, is a public good of transcendent importance."[155]

Notes and References

[1]Scarf, M. (1996) Keeping Secrets. *The New York Times,* June 16, sec. 6, p. 38.

[2]Health Privacy Project, Institute for Health Care Research and Policy, Georgetown University, *The State of Health Privacy: An Uneven Terrain* (July 1999) (Executive Summary, p. 2), available at <http://www.healthprivacy.org/resources/statereports/contents.html> [hereinafter State of Health Privacy].

[3]Skolnick, A. (1998) Opposition to law officers having unfettered access to medical records. *JAMA* **279(4)**, 257–259.

[4]Allen, A. Medical privacy? Forget it! *Med. Econ.* **75(9)**, 151; based on longer article that appeared as Exposed: computer technology, managed health care and genetic science are all undermining the American tradition of medical privacy in the name of progress. What can or should we do about it? *The Washington Post* (Feb. 8, 1998).

[5]Scarf, M. (1999) Brave new world: the threat to privacy that didn't go away. *The New Republic,* July 12, p. 16.

[6]Institute of Medicine, Division of Health Care Services, Committee on Improving the Patient Record (1997) *The Computer-Based Patient Record: An Essential Technology for Health Care* (rev. ed.) (Dick, S., Steen, E. B., and Detmer, D. E. eds.) National Academy Press, Washington, DC.

[7]Blevins, S. (1999) Medical privacy invasion? *The Washington Times,* July 22, p. A17.

[8]A graphic chart of where patient information flows both inside and outside the health care industry is available at the American Health Information Management Association's website, <http://ahbima.org/media/flow_patient.html>, and forms the basis for this discussion.

[9]Borzo, G. (1997) PCASSO with a mouse. *Am. Med. News* (Log On), Oct. 13, available at <http://www.ama-assn.org/sci-pubs/amnews/net_97/logo1013.htm>

[10]Mitka, M. (1999) Weaving webs for physicians. *JAMA* **281(12),** 1070–1071; Winslow, R. (1997) The doctor: as medical sites proliferate on the web, more physicians are shedding their technophobic past. *The Wall Street Journal*, Dec. 8, R10.

[11]Carns, A. (1999) Patients' next choice: whether to keep files stores on the internet. *The Wall Street Journal,* Aug. 16, B1.

[12]California HealthCare Foundation (2000) *Privacy: Report on the Privacy Policies and Practices of Health Web Sites.* January, p. 4; conducted by J. Goldman and Z. Hudson (Health Privacy Project, Georgetown University) and R. M. Smith, available at <http://ehealth.chef.org/priv_po13/index_show.cfm?doc_id=33>.

[13]Coleman, D. L. (1999) Who's guarding medical privacy? *Business & Health.* **17(3),** 29; Fuller, B. P., et al. (1999) Ethics: privacy in genetics research. *Science,* **285,** 1359–1361.

[14]The Center for Public Integrity (1998) *Nothing Sacred: The Politics of Privacy* p. 27, available at <http://www.publicintegrity.org> (hereinafter *Nothing Sacred*); Coleman, supra note 13.

[15]Woodward, B. (1995) Sounding board: the computer-based patient record and confidentiality. *New Engl. J. Med.* **333(21),** 1419–1422.

[16]Upton, J. (1999) U-M medical records end up on web. *The Detroit News,* Feb. 12.

[17](1998) Privacy? At most HMOs you don't have any. *USA Today,* July 13, p. 12A; Bass, A. (1998) HMO puts confidential records on-line. *The Boston Globe,* Mar. 7, p. 1.

[18]Riley, J. (1996) Case study: with old password, cracking the code. *Newsday,* Mar. 31, p. A30.

[19]Woodward, supra note 15, p. 1420.

[20]Coleman, supra note 13.

[21]Nothing Sacred, supra note 14, p. 17.

[22]Riley, J. (1996) Open secrets: changes in technology, health insurance making privacy a thing of the past. *Newsday,* Mar. 31, p. A05.

[23]Scarf, supra note 5.

[24]Woodward, supra note 15, at 1420.

[25]Coleman, supra note 13.

[26]Allen, supra note 4.

[27]Scarf, supra note 1.

[28]Ibid.

[29]O'Harrow, Jr., R. (1998) Between doctor and patient, a three-way relationship: adding pharmacy benefit manager to treatment mix can bring interference, information, and savings. *The Washington Post,* Sept. 17, p. A26. (hereinafter Between Doctor).

[30]Allen, supra note 4.

[31]Breitenstein, A. G. (1998) Let states fill the gap. *USA Today*, April 6, p. 11A; *see also* Allen, supra note 4.

[32](1998) Privacy? At most HMOs you don't have any. *USA Today*, July 13, p. 12A.

[33]Skolnick, supra note 3, p. 257.

[34]Peter D. Hart Research Associates, Shell Poll (May 1999) (available in POLL database, WESTLAW, Queston ID: USHARTSH99MR R23B5, no. 088) (67% of respondents called such sharing a °"major" invasion of privacy, 27% called it a "minor" invasion).

[35]Princeton Survey Research Associates, California HealthCare Foundation survey (Jan. 28, 1999), press release at http://www.chcf.org/press/viewpress.cfm?itemID=362.

[36]The Tarrance Group: Lake, Sosin, Snell, Perry, National Partnership, Family Matters Survey (Feb. 1998) (available in POLL database, WESTLAW, Question ID: USTARR.98FAMM R85, no. 082) (61% for "extremely important," 29% for "very important").

[37]Yankelovich Partners, Inc., Time/CNN/Yankelovich Partners Poll (Feb. 1996) (available in POLL database, WESTLAW, Question ID: USYANK.022396 R28, no. 043).

[38]Louis Harris and Associates, Equifax-Harris Poll (Nov. 1993) (available in POLL database, WESTLAW, Question ID: USHARRIS. 110593 R3C, nos. 008, 010, 015). See also Wirthlin Group (June 1994) (available in POLL database, WESTLAW, Question ID: USWIRTH.94JUNE R10H, no. 011) (most people said they were either "extremely concerned" [55% of respondents] or "somewhat concerned" [27%] about the privacy of their medical records). *See* generally medical privacy opinion polls at <http:www.epic.org/privacy/medical/polls.html>.

[39]Gostin, L. (1997) Health information and the protection of personal privacy: ethical and legal considerations. *Ann. Int. Med.* part 2, **127,** 683–690.

[40]Biddle v. Warren General Hospital, 1998 WL 156997 (Ct. App. 1998), aff'd, 715 N.E.2d 518 (Ohio 1999).

[41]California HealthCare Foundation survey, supra note 35.

[42]Woodward, supra note 15, p. 1421.

[43]*See* generally Gorlin, R. A. ed., *Codes of Professional Responsibility: Ethics Standard in Business, Health and Law*, 4th ed., BNA, Wash-

ington, D.C. 1999; compiling ethical codes and standards for health care professions.

[44]American Medical Association, Code of Medical Ethics, Fundamental Elements of the Patient-Physician Relationship, Element No. 4, as found in Gorlin, supra note 43, p. 342.

[45]Goldstein, A. (1999) Long reach into patients' privacy: new uses of data illustrate potential benefits, hazards. *The Washington Post*, Aug. 23, p. A01.

[46]Lee, T. (1999) Too much privacy is a health hazard. *Newsweek*, Aug. 16, p. 71.

[47]Goldstein, supra note 45.

[48]See generally Furrow, B. R., Greany, T. L., Johnson, S. H., Jost, T. S., and Schwartz, R. L. (1995) *Health Law*, 1, 4–34 (enumerating various mandatory disclosures under state and federal law).

[49]Tarasoff v. Regents of the Univ. of Calif., 551 P.2d 334 (Calif. 1976).

[50]James, F. (1999) Children's Vaccine Registry Raises Medical Privacy Fears. *Chicago Tribune*, May 7, p. 1.

[51]Ibid.

[52]Melton III, L. J. (1997) The threat to medical-records research. *New Engl. J. Med.* 337(2), 1467–1469.

[53]Ibid.

[54]Ibid. *See also* Liesesgang, T. J. Potential effect of authorization bias on medical record research. *Am. J. Ophthalmol.* 128(1), 129 (describing Mayo Clinic study in which authors concluded that "laws requiring written authorization for research use of medical records could result in substantial biases in etiologic and outcome studies").

[55]McCarthy, D. B. et. al. (1999) Medical records and privacy: empirical effects of legislation. *Health Serv. Res.* 34(1), 417–425.

[56]*See* Association of American Medical Colleges, *Confidentiality of Medical Records: AAMC Report on the Minnesota Experience*, available at <http://www.aamc.org/advocacy/issues/research/minrepot.htm> (concluding that the "Minnesota statute poses a number of threats to research and may threaten both health policy and public health objectives").

[57]Melton, supra note 52, p. 1469.

[58]45 C. F. R. § 46.116(d).

[59]U. S. General Accounting Office (1999) Medical records privacy: access needed for health research, but oversight of privacy pro-

tections is limited. GAO/HEHS-99-55, Feb., p. 14, available at <http://www.gao.gov/reports.htm>.

[60]Woodward, B. (1999) Challenges to human subject protections in US medical research. *JAMA* **282(20),** 1947–1952, (hereinafter Challenges); Woodward, B. (1997) Medical record confidentiality and data collection: current dilemmas. *J. Law, Med. Ethics* **25(2&3),** 88–97.

[61]GAO Report 99-55, supra note 59.

[62]*See* generally Allen, supra note 4.

[63]Ignani, K. (1999) Do not retard progress. *USA Today,* July 13, p. 12A (lobbyist for American Association of Health Plans, observing that health progress could not occur without relatively unrestricted information-sharing); Pear, R. (1999) Future bleak for bill to keep health records confidential. *The New York Times,* June 21, p. A12 (insurers and HMOs routinely use information in patient files to contact them concerning health benefits, "to detect fraud, to learn which treatments are most effective, to evaluate the work of individual doctors and to identify those who order too many tests and procedures," and they argue that it is "totally impractical for them to request permission each time they want to use a patient's medical records.") (hereinafter Future Bleak.)

[64]Ignani, supra note 63.

[65]Goldstein, supra note 45.

[66]Allen, supra note 4.

[67]O'Harrow, Jr., R. Plans' access to pharmacy data raises privacy issue: benefit firms delve into patients' records. *The Washington Post,* Sept. 27, p. A01 (hereinafter Plans' Access); O'Harrow, Between Doctor, supra note 29.

[68]O'Harrow, Plans' Access, supra note 67 (reporting GAO estimate of $600 million in prescription cost savings in the federal employees health program in 1995); O'Harrow, Between Doctor, supra note 29 (reporting PBM official estimates of $3 billion in annual drug-cost savings for health plans nationally).

[69]O'Harrow, Plans' Access, supra note 67.

[70]Ibid.

[71]O'Harrow, Jr., R. (1998) Prescription sales, privacy fears: CVS, Giant share customer records with drug marketing firm. *The Washing-*

ton Post, Feb. 15, p. A01 (in the context of two retail drug stores' disclosure of patient prescription information to a database marketing firm, quoting medical ethicist Robert Veatch that "the conflict is so basic it's probably indefensible.") See infra note 82 and accompanying text.

[72]Van Meveren, S. (1999) Don't hinder crime-fighting. *USA Today,* Aug. 20.

[73]Hallam, K. Price of privacy may be too high: FBI, Justice Department say patient confidentiality bills would threaten fraud crackdown. *Modern Healthcare,* May 13, p. 12.

[74]Video Privacy Protection Act, 18 U.S.C.S. § 2710 (warrant, grand jury subpoena, or court order required).

[75]Moore, J. D. (1998) Confidentiality casualty: patient billing printouts released in Kansas fraud case. *Modern Healthcare*, Sept. 14, p. 3; (1998) A trail of mistakes led to breach of patient-data confidentiality. *Modern Healthcare*, Sept. 21, p. 80.

[76]Skolnick, supa note 3.

[77]*State of Health Privacy*, supra note 2 (Executive Summary).

[78]Ibid.

[79]Ibid. (see Georgia summary).

[80]*See* generally, Panzitta, L. J., and Fleming, T. M. (1997) Hospital medical records. Georgia Hospital Law Manual, 4th ed. Georgia Academy of Attorneys, Marietta, GA.

[81]Georgia Code Ann § 33-24-59.4.

[82]*See* O'Harrow, Plans' Access, supra note 67. See also O'Harrow, Jr., R. (1998) CVS also cuts ties to marketing service; like giant, firm cities privacy on prescriptions. *The Washington Post*, Feb. 19, p. E01.

[83]O'Harrow, Plans' Access, supra note 67 (quoting G. D. Lundberg, then-editor of the *Journal of the American Medical Association*).

[84]Palmer, L. D. (1999) Privacy bill for patients is debated. *The Boston Globe,* June 7, p. A1.

[85]Goldstein, supra note 45.

[86]Nothing Sacred, supra note 14, at 5.

[87]5 U.S.C.S. §§ 552a.

[88]5 U.S.C.S. § 552(b)(1)-(9).

[89]Gostin, L. O., (1995) Health information privacy. *Cornell L. Rev.* **80,** 451, 501–503.

[90]42 C.F.R. § 482.24(b)(3) (Medicare Conditions of Participation for Hospitals).

[91]42 U.S.C.S. § 290dd-2; 42 C.F.R. Part 2, §§ 2.1 et seq.

[92]42 U.S.C.S. § 12112(d)(3)-(4).

[93](1999) Privacy Eighth Circuit says ADA confidentiality rules protect all employees, not just disabled. *Health L. Rptr.* **8,** 1410 (BNA), Washington, DC.

[94]Whalen v. Roe, 429 U.S. 589 (1976).

[95]Hodge, Jr., J. G., Gostin, L. O., and Jacobson, P. D. (1999) Legal issues concerning electronic health information: privacy, quality, and liability. *JAMA* **282(15),** 1466–1471.

[96]E.g., Whalen v. Roe, supra note 94 (holding no unconstitutional deprivation of right or liberty by N.Y. statutes that required reporting to state agencies of certain prescription drug information); see Gostin, supra note 89, p. 495–498 (collecting cases that illustrate judicial deference to government's need to acquire and use information about individuals).

[97]Jaffee v. Redmond, 518 U.S. 1 (1996).

[98]P.L. 104-191, § 264(c)(1), see 42 U.S.C.S. § 1320d-2 (notes).

[99]Copies and comparisons of these various bills are available through several websites, including that of the Health Privacy Project of Georgetown University (text of bills) <http://www.healthprivacy. org/legislation/index.shtml>: National Coalition for Patients Rights (legislative scorecards) <http://www.nationalcpr.org/>: American Health Information Management Association (analyses of bills) <http:/www.ahima.org/dc/index.html>.

[100]Standards for Privacy of Individually Identifiable Health Information, 45 C.F.R. Parts 160 and 164, Nov. 3, 1999. The proposed regulations are available at <http://aspe.hhs.gov/admnsimp/ index.htm>, and the pdf version downloaded from this is hereinafter referred to as *HHS Proposed Rule.*

[101]HHS Proposed Rule, supra note 100, pp. 24, 27, 41; (1999) Privacy: Clinton Releases Proposed Regulation, Officials Stress Limits on HHS' Authority. *Health Care Daily Rpt.* Nov. 1. BNA, Washington, DC.

[102]*See* generally Congressional Research Service, Library of Congress, CRS Issue Brief: Medical Records Confidentiality, Oct. 27, 1999 (comparing proposed bills on key controversial issues); National

Committee on Vital Health Statistics, Health Privacy and Confidentiality Recommendations (June 25, 1997) (discussing key issues in context of recommendations to HHS), available at <http://www.ncvhs.hhs.gov/privrecs.htm> (hereinafter NCVHS Recommendations).

[103]Pear, R. (1999) Patient privacy debated; Clinton unveils proposed new medical-records rules. *San Diego Union-Tribune,* Oct. 30, p. A1 (hereafter Pear, Patient Privacy).

[104]American Health Information Management Association, Confidentiality of Medical Records: A Situation Analysis and AHIMA's Position, available at <http://www.ahima.org/infocenter/current/white.paper.html>.

[105]Aston, G. (1999) State laws show mixed results on privacy. *Am. Med. News* **42(30),** 6. (hereinafter State Laws).

[106]Privacy: HIAA passes resolution urging Congress to pass bill protecting health information. *Health Care Daily Rpt.* BNA, Washington, DC, Feb. 26, 1999.

[107]Association of American Medical Colleges, *Issue Briefs: Medical Records Confidentiality* (Position Paper), available at <http://www.aamc.org/advocacy/issues/research/mrcon.htm>(hereinafter AAMC Position Paper).

[108]Aston, State Laws, supra note 105; Aston, G. (1999) Battle lines over bills on medical records privacy. *Am. Med. News* **42(18),** 1 (hereafter Battle Lines).

[109]Aston, G. (1999) Delegates firm up privacy policies. *Am. Med. News* **42(26),** 1.

[110](1999) Privacy: national groups urge congress to rethink preemption of state laws in privacy bills. *Health Care Daily Rpt.,* BNA, Washington, DC, May 12, 1999.

[111]HHS Proposed Rule, supra note 100, pp. 38–39, 326–328.

[112](1999) Privacy: Bennett says medical privacy bill markup may not happen until September, *Health Care Daily Rpt.,* BNA, Washington, DC, June 21, 1999 (hereinafter Bennett Says); Palmer, supra note 84.

[113]HHS Proposed Rule, supra note 100, pp. 27, 365–366; Pear, R. (1999) Clinton to stress medical privacy: regulations on confidentiality of records to be proposed soon. *San Diego Union-Tribune,* Oct. 27, p. A1 (hereinafter Pear, Clinton to Stress).

[114](1999) HHS privacy reg seeks to protect subjects in privately-funded studies. *Health News Daily* **11(210),** Nov. 1, 1999. Page, S. L. and Larios, D. W. (1999) Proposed federal privacy rules: locking the electronic file cabinet. *The Health Lawyer,* **12(2),** 1–10.

[115]Bennet Says, supra note 112; Pear, Future Bleak, supra note 63.

[116]*HHS Proposal Rule,* supra note 100, pp. 75–76.

[117]*AAMC Position Paper,* supra note 107; *AAMC Minnesota Report,* supra note 56.

[118]Ibid.

[119]Aston, G. (1999) Privacy policy will have impact—any way it ends up. *Am. Med. News* (hereinafter Privacy Policy), Nov. 15, 1999.

[120]See Woodward, supra note 60 ("The dislike of some researchers for the consent requirement, which is the key to a research subject remaining a research subject rather than becoming a research object, is well known.").

[121]Aston, Privacy Policy, supra note 119; Privacy: medical records used in research need greater protection, Senate panel hears. *Health Care Daily Rpt.* BNA, Washington, DC, Feb. 25, 1999.

[122]*NCVHS Recommendations,* supra note 102, p. 11.

[123]*HHS Proposed Rule,* supra note 100, pp. 209–228; HHS Privacy Reg. supra note 114.

[124]*HHS Proposed Rule,* supra note 100, pp. 220–223.

[125]*NCVHS Recommendations,* supra note 102, p. 12.

[126](1999) Caregivers walk fine line in aiding police, protecting confidentiality. *Med. Ethics Advisor,* **15(8),** 85–88; Pear, Future Bleak, supra note 63 (AMA's position); Palmer, supra note 84 (American Hospital Association's position).

[127]Testimony of Ronald Weich, on behalf of ACLU, before Senate HELP Committee on "Medical Records Confidentiality in a Changing Health Care Environment," April 27, 1999, available at <http://www.aclu.org/congress/lg042799a.html>.

[128]Pear, Clinton to See, supra note 113.

[129]*HHS Proposed Rule,* supra note 100, pp. 181–194.

[130]Rubin, A. (1999) Privacy initiative elicits praise, concern. *The Los Angeles Times,* Oct. 30, p. A12.

[131]Gemingnani, J. and Rowell, N. (1999) Privacy prompts partisan scuffle: delays in the passing of medical privacy legislation. *Business Health* **17(7),** 8.

[132]Nothing Sacred, supra note 14, pp. 33–37; Aston, Battle Lines, supra 108; Palmer; supra note 84; Pear, Future Bleak, supra note 63.

[133]*HHS Proposed Rule,* supra note 100, pp. 29–30, 34.

[134]Ibid. pp. 69–70, 92–99, Pear, Patient Privacy, supra note 103.

[135]*HHS Proposed Rule,* supra note 100, pp. 33–34, 155.

[136]Ibid p. 101.

[137]Pear, Patient Privacy, supra note 103; Rubin, supra note 130.

[138]*HHS Proposed Rule,* supra note 100, pp. 26, 28–29, 50.

[139]Rubin, supra note 130.

[140]*HHS Proposed Rule,* supra note 100, pp. 25–28, 126–128.

[141]Ibid.

[142]Ibid pp. 31–32, 128–129.

[143]Ibid pp. 137–138.

[144]Pear, Patient Privacy, supra note 103.

[145]*HHS Proposed Rule,* supra note 100, pp. 249–300.

[146]Ibid pp. 36–37, 286–291.

[147]Ibid pp. 30–31, 109–117.

[148]Rubin, A. (1999) Clinton to seek rules on medical record privacy. *Los Angeles Times,* Oct. 29, p. A1.

[149]Hargrove, T. (1999) Health data privacy pushed: new U.S. rules would limit access to medical records. *Chicago Sun-Times,* Oct. 30, p. 1.

[150]*HHS Proposed Rule,* supra note 100, pp. 38, 301–326.

[151]Olmstead v. United States, 277 U.S. 438, 479 (1928) (Brandeis, J., dissenting).

[152]Allen, supra note 4.

[153]Palmer, supra note 84.

[151]Olmstead v. United States, 277 U.S. 438, 479 (1928) (Brandeis, J., dissenting).

[152]Allen, supra note 4.

[153]Palmer, supra note 84.

[154]Goldman, J. (1998) Protecting privacy to improve health care. *Health Affairs* **17(6),** 47–60.

[155]Jaffee v. Redmond, 518 U.S. 1, 11 (1996).

Abstract

Bioethicists typically justify medical privacy with the value of autonomy. Although this justification highlights an important concern in privacy issues, it is insufficient for some of the most important and controversial cases of medical privacy. The central problem with autonomy-based justifications is that protecting patient privacy can sometimes undermine the autonomy of third parties. Consequently, if we are to justify medical privacy we cannot limit our concern to the effect privacy has on autonomy. A plausible alternative to a justification that focuses only on autonomy is one that includes the value of equality. The ideal of equality can help us better understand why medical privacy is important and how we can weigh its importance against other conflicting social values.

An Egalitarian Justification of Medical Privacy

Patrick Boleyn-Fitzgerald

I initially developed an interest in medical privacy when I served on the institutional review board for Louisiana's Department of Health and Hospitals. Our board reviewed only research protocols that occurred in Department of Health and Hospitals institutions, and it often reviewed protocols that had already been reviewed by another institutional review board—sometimes several other institutional review boards. At this point in the review process, there were rarely any striking ethical problems. But one issue did continually arise in our deliberations: the confidentiality of Medicaid information.

I must admit that at first, I wasn't sure that the goal was worth any effort at all, and after a couple of six hour meetings, I seriously questioned whether it was worth the effort we were putting into it. Did we really need to spend hours discussing how to minimize the number of individuals who knew that a patient was on Medicaid? What reasons did we have to go to such lengths to keep Medicaid information confidential?

My initial cynicism with medicine's preoccupation with privacy has not led me to reject the concern. It has led me, however, to believe that the reasons we should be concerned about

privacy are often not the ones articulated in the bioethics literature. Bioethicists usually argue that privacy is important because
it protects patient autonomy. A bare concern for autonomy, however, is inadequate to justify privacy in important instances. Instead
of autonomy, our legitimate concerns in privacy often lie in our
goal of creating an equal society.

Why Do People Want Privacy?

This chapter does not focus on all cases where we think
privacy is important. Indeed, that may be impossible, as some
have argued that privacy is undefinable. Sonia Le Bris and Bartha
Maria Knoppers recently claimed that "although privacy is a
commonly used and frequently invoked concept, it is multifac-
eted, fluid, and evolving. Any attempt to circumscribe the concept in a uniform and monolithic manner is inappropriate. Indeed,
such an approach would undermine its very richness, fluidity,
and malleability . . ." Regardless of whether we can give a final
analysis of privacy, however, all cases of privacy do involve a
kind of secrecy. Privacy limits the information others may have.
The focus of this chapter lies primarily with cases where agents
want this secrecy for a specific reason. An individual may want
to keep information secret because he wants to affect the way
others act. This is a frequent motivation, and it seems to explain
the most widely discussed cases where advocates of privacy
worry about its erosion. A patient may not want his employer
knowing about a medical condition because he is afraid of
losing his job. A patient may not want an insurance company
knowing about a medical condition because he may be denied
insurance. A patient may not want his sexual partner knowing
about a sexually transmitted disease because she may leave
him. A patient may not want the police knowing about his
medical condition because it will give them evidence that he has
committed a crime. Hence, one reason a patient may want pri-

vacy is that it may prevent a third party from acting in a way that is contrary to the patient's interests.

I will focus primarily on cases where this is the motivation for privacy, but I will also touch on a second, related motivation. A second reason a patient may want information kept private is that privacy may protect her from social judgment. Members of a society may judge others for a wide variety of activities or personal characteristics, and an individual may want to avoid the social judgement. This motive overlaps significantly with the first. The desire to avoid negative social judgment may be closely tied to the desire to prevent others from treating one in a specific manner. So a criminal may want to avoid both criminal punishment and being seen as a criminal. The patient with a sexually transmitted disease may want to avoid both the loss of his marriage and the appearance that he committed adultery. Sometimes, however, the desire is independent. A patient may want to hide information merely because it will change the way others think about him. So a patient may not want others to find out about his illness because that knowledge will lead others to pity him, or perhaps even look at him as disgusting or repulsive. Confidentiality provides a space of safety where individuals need not be afraid that such information will be disclosed.

Justifications for Privacy

Now we have an idea of why some patients may want to keep information private. But why should medical professionals work to keep information private? What is the justification? Some theorists argue that when professionals keep information private, it promotes effective medical treatment. Keeping confidences promotes trust and the willingness on the part of the patient to disclose fully information to a doctor. A second argument is that keeping a promise or confidence is indicated by justice. When patients explicitly request information to be kept private, or when they reasonably assume that information will be kept private, then the medical profession has an obligation to keep the actual or implicit promise. To break that promise is to fail to give individuals their due, and hence to treat them unjustly.

Many—if not most—ethicists, however, discount these consequential and juridical arguments, and ground our concern for privacy in the value of autonomy. Sissela Bok, for example, argues that autonomy is fundamental in the argument for privacy. She states, "The first and fundamental premise is that of individual autonomy over personal information. It asks that we respect individuals as capable of having secrets. Without some control over secrecy and openness about themselves, their thoughts and plans, their actions, and in part their property, people could neither maintain privacy nor guard against danger."[2] Tom Beauchamp and James Childress make the clearest claim in favor of an autonomy-based justification. They argue that "the primary justification resides in a . . . rationale based on the principle of respect for autonomy."[3] They also suggest that the courts employ the same argument, that the legal right to privacy is "primarily a right of self-determination in decision-making."[4] In the bioethics literature, then, autonomy is thought central to the justification of privacy practices. But is the desire to protect autonomy an adequate justification for the protection of privacy?

Problems with Autonomy

Autonomy is inadequate as a justification for privacy when the motivation for privacy is a desire to affect the actions of others. The problem in these cases is that protecting the autonomy of a patient through the practice of protecting privacy can undermine the autonomy of the third parties whom the patient wants to affect. Indeed, the desire to keep information private may be grounded in a desire that others not be able to make an autonomous choice.

Consider, for example, the case of a husband who has tested positive for a treatable sexually transmitted disease. The usual practice is to keep this information private as long as doing so does not pose a serious risk of harm to the spouse. It sounds suspect, however, to say that we do this for reasons of autonomy, because we restrict the autonomy of someone regardless of what choice we make. If we disclose the information, then we restrict the autonomy of the patient to control information about his private life. If we fail to disclose the information, however, then we restrict the autonomy of the spouse to make an autonomous choice about continuing her marriage. If we are only worried about individuals making autonomous choices, then it is not clear that we should keep the information private.

The same kind of conflict seems to exist in other cases. If a patient asks us not to disclose information to a life insurance agent we preserve the autonomy of the patient by not disclosing it. If, however, the life-insurance agent thinks that she has full medical information when she decides whether to insure the patient, then our failure to disclose has prevented her from making an autonomous choice. Concern for autonomy alone is not enough to help us distinguish here. The same problem would occur for a prospective employer who is trying to decide whether to hire the patient. Protecting privacy gives the patient self-determination over disclosure, but it also undermines the ability of a third party to make a choice with substantial understanding.

Some might respond that we do not infringe on the autonomy of the spouse, the insurance agent, or the employer because each of them can still make a choice when we keep information private. On the other hand, the decision to disclose would deny the patient a choice of what to do with her private information. It is true that our decision to disclose would deny the patient a choice, and it is also true that our decision not to disclose would not deny any third party a choice. Having a choice, however, is only part of what it means to be autonomous. If choice were a sufficient condition for autonomy, then we would be content with patient consent rather than informed consent. Hence Beauchamp and Childress state, "We analyze autonomous action in terms of normal choosers who act (1) intentionally, (2) with understanding, and (3) without controlling influences that determine their action."[5] In the three aforementioned cases the decision not to disclose undermines the second condition Beauchamp and Childress state is necessary for autonomous choice: understanding. The spouse is not informed about an important aspect of her marriage. The insurance agent is not informed about a piece of information essential for determining risk. The employer is denied an important piece of information for determining the best candidate for the job. These agents do not act with understanding.

We can highlight this issue by imagining a patient who does not know the potential negative consequences of disclosing information to a prospective employer or insurance agent. Do we need to make sure that the patient's decision to release private information is informed? When we ask the patient "May we release this information to your employer?" and they answer "Yes" do we need to add, "You realize that these are the following risks of releasing such information: . . ."? Whichever way we answer this question would be problematic for an autonomy-based justification. If we maintain that we do not need to inform patients about the risks of releasing their private information, then our getting consent doesn't really protect autonomy. Uninformed consent for the release of private information does no more to protect autonomy than uninformed consent for an operation. In neither case does the individual have enough information to make an authentic choice. If, on the other hand, we maintain that we do need to inform the patient of these risks, then our concern about autonomy centers on individuals making informed choices. This, however, is a problem for the defense of privacy, because privacy undermines informed choices for third parties. A mere concern for autonomy does not suggest that we should keep information private.

If the defense of privacy is to center on autonomy at all, it must somehow differentiate the autonomy privacy protects from the autonomy it undermines. We have to say more than that we should respect or promote autonomy. One way to accomplish this would be to move away from talk about autonomy and move toward talking about rights. Both disclosing and not disclosing respects the autonomy of some and undermines the autonomy of others, but one might claim that the patient has a *right* to control the private information, whereas third parties have no right to receive it. This would be an example of a deontological justification for respect of privacy, but it runs into some serious difficulties. The concept is a term that has been wielded without much discrimination over the past 50 years. This is not to say that the concept of a moral right is not useful in moral discourse, but it probably is not very useful as a justification of moral claims. The concept of a moral right is very useful in describing a certain kind of moral relationship—a relationship where one individual has a claim against another—but this relationship is the conclusion of a moral argument, not a premise.[6] In other words, it is not enough to say that patients have rights to control their private information; we must also say why we believe that they have such rights. We can see this intuitively with the husband who wants to keep his sexually transmitted disease private. Some might say that the wife has "a right to know." At this point, we would have to say why she has this right in this specific instance. Ironically, a common justification for individual rights is the value of autonomy. An individual may have an interest in autonomous action that is protected by the ascription of a right, but of course the weakness of autonomy-based justifications is what led us to look to moral rights in the first place. So it is not enough to say that patients have rights to their private information while others do not. That statement is really nothing more than a restatement of the conclusion that we have been scrutinizing. What we want to know is whether there is any justification for the rights we commonly ascribe.

An alternative way to justify privacy is to look to the importance of patient autonomy over private information vs the importance of third party autonomy over a patient's private information. If we can say that respecting or promoting patient autonomy is more important than the autonomous choices by third parties, then we can provide a justification for privacy practices even in cases where it may undermine the authentic choice of a third party. Is there any reason to think that patient autonomy in these cases is more important than the autonomy of third parties? Does patient autonomy serve any special social function? I think that it does, and that the explanation lies in the relationship between patient autonomy and the social goal of equality.

An Egalitarian Justification

Why do we think it important to keep some personal information private, but do not hesitate to publicize other kinds of personal information? Why do we try to protect Medicaid information, but publicize the recipients of National Endowment for the Humanities fellowships? Both are receiving grants from the government, yet only one of them raises concerns about privacy. Why does the Nobel Prize Committee fail to get its recipients to sign an informed consent form before it releases their name to the press? We might think that the difference lies in the fact that publicity about winning the Nobel Prize presents no risks to the recipients, but that would be wrong. Recipients are thrust in front of the public eye, and this may radically change their lifestyle. Individuals who value a reclusive life may suffer because of this. Of course, most individuals would consider the fame from a Nobel Prize something good, but if we were really concerned about a prize winner's autonomy, we would have to keep an eye out for those who did not. So autonomy cannot really explain the differences in treatment here, but I think the ideal of equality might be able to give us some insights.

The ideal of equality is used in a variety of ways, so I should clarify how I am using the term. Many contemporary theories of equality focus on cosmic injustice and try to minimize its effects. Elizabeth Anderson has called these theories "luck egalitarianism" because they aim at minimizing the effects of brute bad luck. This is not what I mean by equality. When I speak of a principle of equality, I am speaking of a theory that is both empirical and normative. Ideal equality describes both facts about who we are, and ideal social relationships that reflect those facts. In other words, we are equal and our policies and attitudes should reflect that equality. When I say that we are equal, I am not saying that we are the same in every respect, but I am saying that we are the same in *some* respects and that the ways in which we are the same are *important*. They are so important, they should have a significant impact on how we relate to each other.

In what ways are we equal? We all face moral choices and the responsibility for those choices. We all have basic needs that must be satisfied if we are to function as human beings, as productive members of society, and as citizens in a democratic state. We all have a limited life span. We are all subject to pain and suffering. And none of us will satisfy all of our desires. These are common features of humanity. When we see someone in terms of these common features, we relate to their basic humanity; we relate to them as equals. The principle of equality suggests that in many cases, we should view others primarily in terms of their humanity. We should look at others as equals, and our policies should reflect that presumption of equality.

Often, however, individuals in society do not look at each other as equals. This is not always a problem for the ideal of equality. For practical reasons, we sometimes have to focus on our differences. We have to make a distinction between a doctor and a nurse; we have to make a distinction between a specialist in cardiology and a specialist in dermatology; and we have to make distinctions between patients, because they each need different treatments. But this is in line with the principle of equality.

We are not the same in all respects, so we should not be treated the same in all respects. We are, however, the same in some respects, and the principle of equality says that those similarities are important. We have a problem from the perspective of ideal equality when our differences are given too much weight, and when our similarities—the ways in which we are empirically equal—are underappreciated.

This is a very quick sketch of a relational theory of equality. When we understand the ideal of equality as an ideal social relationship—an ideal of individuals relating to each other primarily according to their common humanity—then we can see that in at least two areas, privacy serves as a tool to promote ideal equality.

The first area covers cases where the disclosure of medical information might stigmatize patients. Consider a common privacy concern in medicine: the disclosure of a patient's HIV status. The ideal of equality suggests that we should relate to those who are HIV+ as equals. We should see their common humanity. Often, however, this is not the case. Patients with HIV often suffer social harms in addition to the physical harms caused by their illness. Others often see them with pity or disgust. Privacy acts to prevent this kind of relationship. By hiding the medical information, the medical profession prevents at least some instances where the patient will be judged an inferior or be dehumanized by others who know his condition.

The second area covers issues in the distribution of medicine. Medicine is something humans need. We are all vulnerable to conditions where the only thing that stands between us and death is medical care. Relational egalitarianism maintains that this empirical equality is normatively important. For reasons of equality, the distribution of medicine should be determined in large part, if not completely, by medical need. Of course, equality is not the only political ideal, and efficiency may lead us to distribute medicine in part by the ability to pay, but this issue is beyond the scope of this paper. We should note, though, that the existence of other political ideals that compete with equality merely suggests that the ideal of equality is not an absolute social value, not that it is unable to give us practical guidance. We should also note that ability to pay is not the only impediment that may prevent medicine from being distributed by medical need. Patients may not seek medical care because they fear that in doing so, they may be exposed to a social harm, such as criminal punishment, social condemnation, loss of life insurance, or loss of a job. Consequently, even if we were to believe that the distribution of medicine should be determined in part by ability to pay, we may have serious problems with other ways in which the distribution of medicine by medical need gets undermined. Because we want medicine to be distributed by need, one goal of the medical profession must be to create a space where patients need not fear resulting social harm.

Another goal must be to create a space where status based on income, birth, fame, and power can be ignored. Creating such a space minimizes the chance of medicine being distributed for illegitimate reasons. Consequently, medicine should strive to be a place where it is possible for professionals to see patients—and where patients can feel secure that they are seen—in terms of their humanity, their equal vulnerability to disease, the ravages of age, and the unavoidable end of death.

The goal of medicine to create a space of equality, and privacy's usefulness in promoting such a space, is helpful in explaining the previous cases where autonomy posed a difficulty. Why would it be important to keep the secret of a husband who has a sexually transmitted disease, or refrain from disclosing information to an insurance company or to an employer? One reason we have not to disclose, in addition to the patient's autonomy, is that respecting privacy contributes to a safe space for patients. A patient need not worry that seeing a doctor will result in harm to their personal relationships, a loss of insurance, or denial of a career opportunity. Respect for privacy promotes both the goal of medicine as a safe space and the social goal of equality.

We can see the same issue at work with the question of disclosing information to law-enforcement agencies. Again we might be tempted to describe this issue in terms of autonomy: that we want criminals to have the autonomy to hide their criminal activities. I think, however, that there is a better explanation. It is not that we think criminals have the right to keep their crimes secret, but rather that we think everyone should feel that they can secure health care without worrying about being punished as a result. Of course, we also want the just prosecution of criminals, so we have a conflict in basic values in these cases. But it is important that we understand what the conflict is so that we can assess how we should adjudicate it. The most important reason we want privacy for criminals is because we believe that even criminals, by virtue of their basic humanity, should feel secure enough to seek medical treatment.

Practical Implications

Thus far I have argued that at least in cases where individuals want privacy in order to affect the decisions of others, it is better to understand privacy in terms of the ideal of social equality rather than individual autonomy. Privacy can help counterbalance social forces that create inequality. Is this a practically important conclusion? Does this conclusion alter what we think about privacy policy, or does it just give us a new reason to do what we have always done?

I think it has two important practical implications. First, this account of privacy suggests that there are at least some cases where it is merely instrumentally important; in other words, it is a means to promote social equality. In these cases, we may find that there are better means to promote our goal. The most obvious case is privacy and health insurance companies. If we see privacy as a means of promoting security in the provision of medicine and thus, promoting the social goal of equality, then we can see that it does a remarkably poor job. Patients must consent to give their medical information to a health insurance company before it is released, but they must release the information if they want insurance companies to reimburse their expenses. Privacy provides no real security here. If, however, medical care was provided publicly, then privacy would be irrelevant. Releasing information to a government agency that pays for medical care is no more problematic (as long as the agency does not release the information to anyone else) than a hospital policy that releases information to the billing department. If privacy of one's medical condition from a health care funding organization is meant to promote social equality, we could address the problem much more effectively by replacing private medical insurance with a public system. To put the point in another way, private insurance is at odds with one of the goals of medicine, namely that patients should not have to worry about whether they will be denied access to medical care. We can reduce this conflict within a private payer system by granting patients privacy rights, but this is not a very effective solution. In general, then, when we see privacy as promoting social equality, we may conclude that certain privacy protections would be unnecessary if we were to make systematic changes in the way medicine is provided.

A second practical implication of grounding some privacy issues in the goal of equality occurs when privacy cannot be secured. There are many medical conditions that stigmatize individuals and that individuals would keep secret if it were possible. Obesity, unplanned pregnancy, full-blown AIDS, and other conditions may be ones that patients wish they could keep private, but cannot. If we are worried about privacy for reasons of autonomy, then these cases pose no special concern. If we focus on equality, however, these cases call out for the attention of the medical community. Social stigma works against medicine's goal to create a space where individuals see each other in terms of their basic humanity. When the tool of privacy fails to secure this goal, the medical community should look to promote the goal of equality in other ways. This may be difficult to achieve, and any progress may require creative solutions, but it is important for the medical community to keep in mind that limiting the effects of stigma is the real concern.

Conclusion

I began this chapter with an account of my skepticism about the value of efforts to keep Medicaid information private. I am less skeptical now than I was during the darkest moments of six-hour long institutional review board meetings. Privacy of Medicaid information is important, in part because Medicaid recipients are sometimes stigmatized. Privacy can minimize that negative consequence. While our concern about Medicaid's stigma can justify the protection of privacy it would certainly be better to replace Medicaid with a system that did not stigmatize. If we focus too narrowly on patient autonomy and patient rights, we may miss the most important role that privacy can play.

References

[1]Le Bris, S. and Knoppers, B. M. (1997) International and comparative concepts of privacy, in *Genetic Secrets: Protecting Privacy and Confidentiality in the Genetic Era*, Rothstein, M., ed., Yale University Press, New Haven, CT, p. 438.
[2]Bok, S. (1989) *Secrets: On the Ethics of Concealment and Revelation.* Random House, New York, NY, p. 120.
[3]Beauchamp, T. and Childress, J. (1994) *Principles of Biomedical Ethics*, 4th ed., Oxford University Press, New York, NY, p. 410.
[4]Ibid., p. 410.
[5]Ibid., p. 123.
[6]Buchanan, A. (1989) Justice and charity. *Ethics* **97,** 558–575.
[7]Anderson, E. (1999) What is the point of equality? *Ethics* **109,** 287–337.

Abstract

In 1996, the health-care industry spent $10 to $15 billion on information technology. This figure is expected to grow, as computer technology is used to implement the computerized patient record, electronic claims and billing, and health information networks. The changing health care landscape during the past decade has encouraged an increased reliance on such technologies.

As our health-care delivery system has shifted from private doctor's offices to large, integrated delivery systems and managed care settings, medical information is now routinely shared among many authorized users. Furthermore, there's a growing cry for this information to flow freely among third-party payers, government agencies, and employers, for purposes far removed from direct patient care. The era of "data driven medicine"[1] has arrived, and patients can no longer count on having a confidential, trusting relationship with their doctors.

Technology, particularly the computer-based patient record, offers increased access to medical information. However, this computerization of medicine has not been accompanied by thoughtful, pro-patient public policy. As a result, we face many new ethical challenges, not the least of which is the severe erosion of patient privacy and confidentiality.

This chapter confronts the ethical implications of our technological era and puts forth recommendations for patient-centered policy solutions to guide public policy. The core components of strong, patient-centered policy include a federal floor (not ceiling) of protection, informed non-coerced patient consent for most disclosures of health information, the right to decline use of a unique patient identifier or linked electronic record system, and policies to limit the information disclosed to payers. Uses of infor-

mation for laudable purposes, such as research and public-health surveillance, should be guided by either patient consent or a suggested model of prospective patient consent delegated to an appropriate review board. Most important, patients' rights to privacy in their relationship with their health-care professional needs to be explicitly acknowledged and legally protected.

Medical Privacy in the Information Age

Ethical Issues, Policy Solutions

Margo P. Goldman

Introduction

We are in the Information Age, and Medicine is no exception. In 1996, the health-care industry spent $10–15 billion on information technology. This figure is expected to grow, as computer technology is used to implement the computerized patient record, electronic claims and billing, and health information networks. The changing health-care landscape during the past decade has encouraged an increased reliance on such technologies.

As our health-care delivery system has shifted from private doctor's offices to large, integrated delivery systems and managed care settings, medical information is now routinely shared among many authorized users. Furthermore, there's a growing cry for this information to flow freely among third-party payers, government agencies, and employers, for purposes far removed from direct patient care. The era of "data-driven medicine"[1] has arrived, and patients can no longer count on having a confidential, trusting relationship with their doctors.

Technology, particularly the computer-based patient record, offers increased access to medical information, and no one wants to be denied: not the police, not the marketers and certainly not the insurers who envision the "health plan–patient relationship" replacing the centuries-old doctor–patient bond. However, this computerization of medicine has not been accompanied by thoughtful, pro-patient public policy. As a result, we face many new ethical challenges, not the least of which is the severe erosion of patient privacy and confidentiality. This loss of privacy will harm the doctor–patient relationship and change the very nature of health care. It is therefore imperative to confront the ethical implications of our technological era and to develop sound, patient-centered policy solutions.

The Current Health-Care Climate

During the last decade, we have witnessed a dramatic shift in the nation's health-care landscape. Spiraling health care costs have given rise to aggressive cost containment measures via insurer-dominated managed care. In order to cut spending, payers review patient information to determine whether to pay for medical services. Such "case management" has required increased amounts of detailed clinical information, sometimes as much as the entire patient record. Blanket waivers signed at the time of health plan enrollment or the start of treatment have allowed this to occur legally. In addition, doctors signing health plan provider contracts are frequently required to waive their patients' privacy rights by agreeing to disclose medical records to the managed-care company. However, patients are usually unaware that their doctors often have to divulge confidential patient information in order to be paid or to receive insurance-company approval for certain treatments.

Second, economic survival for doctors and hospitals has necessitated a "merger mania" among health care facilities and professionals. As a result, we have experienced the decline of freestanding community hospitals and small private practices. Large, integrated delivery systems have become the norm, and with the increased computerization of patient records, patient information has been networked and shared. Statewide networks have been implemented, with 37 states now requiring the collection of detailed information from hospitals.[2]

The increased availability and ease of access to medical records has created a virtual feeding frenzy. The medical record, originally intended to document and facilitate patient care, has become a source of information for both commercial gain as well as some worthwhile purposes (i.e., research and public health). The absence of federal laws and a mere patchwork of state laws have allowed this unfettered access to occur, in stark contrast to people's expectations of confidentiality when they seek medical care. Consider some examples of what is actually happening:

- CVS, a large pharmacy chain, sold prescription information to a marketing firm to conduct a direct mail campaign to promote additional products.[3]
- A man sought evaluation at a sexual dysfunction clinic for impaired potency and subsequently received numerous mailings about commercial remedies. Had he returned for medical treatment, his health insurer may have required his doctor to submit an "Erectile Dysfunction Medical Necessity Form" (*see* Fig. 1). His outrage and humiliation prevented him from seeking additional care (*see* National CPR web site).[4]
- A woman's pregnancy test was positive, but she subsequently miscarried. At her approximate due date, she was devastated to receive mailings from formula and baby-product companies.[5]
- A man applied for long-term care insurance, and learned from his psychiatrist that the prospective insurer urgently demanded a copy of his entire psychotherapy record.[6]

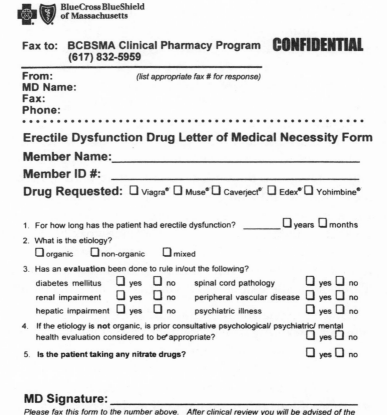

Fig. 1. Erectile Dysfunction Medical Necessity Form

- A man sought drug-abuse treatment, and inquired about having insurance coverage for the services. The clinic checked his eligibility for benefits, and when he reported for work the following day, his employer had been informed of his inquiry. He was immediately terminated from the job, despite satisfactory performance reviews.[7] This is consistent with the fact that 35% of Fortune 500 companies have admitted to using employee medical records in employment decisions.[8]

- A state department of public health plans to study asthma in children by reviewing their school health records without parental knowledge or consent.[9]

This widespread dissemination and use of patients' information is occurring despite an apparent consensus about the importance of confidentiality in health care. However, we are no longer able to protect our patients' privacy with any certainty. The Institute of Medicine accurately predicted this, as follows:

> "an exhaustive list of patient record users would essentially parallel a list of the individuals and organizations associated directly or indirectly with the provision of health care" and that "the fully array of patient record users and the respective needs of each were too extensive .... to address fully."[10]

Because all the potential users of a health information infrastructure have claimed a compelling "need to know," it is likely policymakers will authorize a wide circle of individuals and institutions to access patient records. Large portions of the patient record are *already* being sought for purposes ranging from clinical care, oversight, and cost-containment, making it difficult to imagine how (under these conditions) patient privacy can be meaningfully preserved.

In order to accommodate these expanded uses of the patient record, confidentiality has been redefined. Confidentiality has been understood for centuries to mean what a patient told his or her doctor would not be disclosed to anyone else without the patient's permission. That is no longer the case. In the National Research Council's *For the Record*, confidentiality is redefined as "a condition in which information is shared or released in a controlled manner".[11] Confidentiality is now treated as a tradable commodity, in which patients' privacy rights are balanced against the potential gains of disclosing/using their personal information.

The impact of this is staggering. A recent survey conducted by the California Health Care Foundation found that 15% of all adults surveyed had "done something out of the ordinary" to protect their health-care privacy. This included delaying or avoiding needed treatment, paying out-of-pocket for care, withholding information, or asking the doctor to not include certain information in the record.[12] Another study of psychotherapy in managed care settings found that patients were less likely to be forthright with therapists under a managed-care regimen than in a standard fee-for-service setting.[13]

As people change their behavior to preserve privacy, medical records will be incomplete and inaccurate, as well as unreliable for sound clinical care, research, and public health. At its worst, delaying or avoiding care will undermine early diagnosis, prevention, and intervention. Quality will decline and costs will rise. Because of these deleterious consequences, we must reaffirm the ethical and practical rationales for preserving patients' privacy.

Ethical Basis for Privacy and Confidentiality in Health Care

Early in medical training, students are taught about the importance of the doctor–patient relationship. A noted Harvard Medical School internist, Dr. John Stoekle has said,

"If patients can't rely on confidentiality, then they can't tell us the full story of their illnesses, and the healing process can't begin."[14]

The main purpose of information gathering from a patient is for clinical diagnosis and treatment, and trust in the doctor–patient relationship allows this to occur. The US Supreme Court upheld the importance of absolute confidentiality between psychotherapists and patients in the 1996 Jaffe v Redmond decision. The Court held that without the promise of such communications being private, psychotherapy would be impossible.[15] Second, a 1997 study demonstrated that if adolescents are told that their confidentiality is unconditionally guaranteed, they are much more likely to seek care and communicate openly with their doctors.[16]

In fact, the medical history is generally acknowledged to be crucial for accurate, timely diagnosis. More than half of the information necessary for diagnosis comes from the history provided by the patient. This provides a compelling practical basis for ensuring the confidentiality and privacy of patients' health information.[17]

In addition, patients seek care with the understanding and expectation of respect for their autonomy and control. The US health-care system has a long tradition of requiring informed consent for treatment and research, and allowing people to refuse treatment even if against medical advice. Similarly, patients reasonably anticipate that the information confided to their physician will be used for their medical benefit. Herein lies the intrinsic value of medical privacy and confidentiality: ensuring that peoples' dignity and autonomy are not compromised in the course of their health care. Such ethical tenets are not tradable, and cannot be balanced against other countervailing interests.[18] This is aptly described by Sherri Alpert in *Ethics, Computing and Medicine:*

> "Patients tend to expect that their communications with physicians, nurses, and other health care providers are and will remain confidential. Because patients may presume that such communications have strong legal protection, they generally feel comfortable providing intimate details to advance their medical treatment. Patients are not likely to disclose these details freely unless they are certain that no one else, not directly involved in their care, will learn of them."[19]

These clear reasons for restoring health care privacy and confidentiality, juxtaposed against powerful forces that threaten it, have created a national crisis. The questions we must consider are these:

- How can we allow sufficient access to medical information to facilitate diagnosis and treatment, while still respect-
  ing/protecting privacy?
- In a health-care system largely financed by third parties,
  how can policy guide their access to patient information
  while preserving a confidential clinical relationship?
- Can we ensure the continuation of valuable, needed research
  and public-health initiatives without trampling on people's
  rights?
- How can we translate these concerns to patient-centered
  principles that guide policy implementation?
- What, if anything, is compelling enough to override patients'
  privacy?

Before discussing patient-centered policy recommendations
that address these questions, a review of recent government proposals will be provided.

Federal Legislative Developments

The issue of medical-records privacy appeared on the federal legislative agenda in 1993 and 1994, when Congress considered numerous health care reform proposals. Attached to
every proposal was an amendment, sometimes titled "Administrative Simplification," that laid the groundwork for a vast electronic medical information infrastructure.[20] Weak privacy
provisions were included in these proposals, but the public was
largely unaware of what our government planned for the health
care system. Then, in 1995, when Sen. Bennett (R-Utah) introduced his "Medical Records Confidentiality Act," a front-page
New York Times story disclosed the truth behind the proposal:
The *Times* quoted Boston University health law professor George
Annas, who dubbed it the "Data Bank Efficiency Act of 1995."
He went on to say, "this is certainly not a privacy statute."[21]

Sen. Bennett's bill was not enacted and no significant action occurred until August 1996, when the Health Insurance Portability and Accountability Act (HIPAA) became law. In an eleventh hour conference committee compromise, the Administrative Simplification amendment was attached to HIPAA, which accomplished what earlier proposals had attempted. Administrative Simplification required:

- The assignment of a unique patient identifier to every individual.
- The translation of health information into standard code sets.
- The electronic exchange of health information in standard code sets.
- The mandatory exchange of information among health plans as long as data adheres to uniform standards.
- That anyone who cannot computerize the data must hand it over to a clearinghouse, which will standardize the data.[22]

With its passage, Administrative Simplification was hailed by industry as representing "biggest government initiated change in health care delivery since the introduction of Medicare."[23] A cradle-to-grave medical record on every individual would now be possible, and with it the demise of the confidential doctor–patient relationship.

To their credit, the legislators recognized the privacy implications of a national medical infrastructure and they included a provision in HIPAA that required Congress to enact medical-privacy legislation by August 21, 1999. If Congress failed to act by the deadline, the law specified that the US Department of Health and Human Services (HHS) would promulgate regulations for the privacy of medical information.

During the past two legislative sessions (105th and 106th Congress), numerous confidentiality bills have been introduced and debated. However, most of these proposals have strong industry backing and actually erode privacy by codifying widespread access to patient information. The most patient-centered privacy-protective bills in the 106th Congress were S. 537, introduced by Sen. Patrick Leahy (D-VT), Sen. Edward Kennedy (D-MA) in the Senate, and H.R. 1057 (an identical companion bill) by Rep. Edward Markey (D-MA) in the House. The bills' strength was the acknowledgment of the patient's right to privacy by taking important steps to restrict massive sharing of information. These bills also provided patients with the opportunity to authorize disclosures for care, payment, and other purposes, with the "other purposes" requiring a separate, non-coerced consent. Though the authorization for care and payment was compulsory (patients could not obtain care or payment without signing it), it allowed patients to decline to disclose information if they pay for care themselves. It called for a Federal floor (not ceiling) of protection by not preempting stronger State laws, and extended Institutional Review Board (IRB) oversight to include non-federally funded research.

One of the many access-oriented bills was Sen. Bennett's (R-UT) S. 881. This proposal coerced patient consent for disclosure of information for care, payment, and broadly defined healthcare operations, and contained multiple other exceptions for uses/disclosures *without* patients' consent. This bill fully preempted stronger state law, stripping states of current and future more stringent protections. Sen. James Jeffords' (R-VT) bill (S. 578) and the Health Education Labor and Pension (HELP) Committee bills were similar, save for the degree of Federal preemption of stronger State law.

Other proposals, such as Rep. Gary Condit's (D-CA) H.R. 1941, Rep. Christopher Shays' (R-CT) H.R. 2455, and Rep. Jim Greenwood's (R-PA) H.R. 2470, omitted patient consent by establishing a "statutory authorization" for disclosures/uses of health information.[24] The only clear exception would have been for commercial or marketing purposes, and even these could be circumvented under the guise of "disease management" or pharma-ceutical benefits management. These proposals were just as egregious as those in the Senate, except for the Leahy/Kennedy bill. (Refer to Table 1 for comparison of bills in 106th Congress.)[25]

Congress failed to meet its self-imposed August 21, 1999 deadline, opening the door for the Department of Health and Human Services to issue regulations by February 2000. These were released early in November 1999, to be finalized by February 2000, and implemented in 24–36 months. Though the proposed rules would establish a federal floor of privacy protection, and provide stronger protections for narrowly defined psychotherapy notes, they failed to acknowledge or establish a right to medical privacy. The proposal did away with the time-honored practice of informed patient consent for release of records by allowing this without patients' authorization for care, payment, health-care operations, and multiple other "national priorities." These proposed regulations dramatically shift control away from patients and direct it toward covered entities (providers, plans, and health information clearinghouses).[26] The unavoidable consequence of these regulations.... is that they do not restore the public's confidence that their personal medical information will be kept private."[27]

Patient-Centered Recommendations to Guide Policy

How Can We Allow Sufficient Access to Medical Information to Facilitate Diagnosis and Treatment, While Still Respecting/Protecting Privacy?

In recent years, proponents of the electronic patient record have argued that rapid easy access to medical information will increase the efficiency and quality of patient care. Furthermore, the linkage of patient information to create a longitudinal record could transcend geographical and physical barriers. A patient from Massachusetts presenting to a California emergency ward

Table 1. Legislative Scorecard

	Aims to Protect Privacy by Restricting Widespread Sharing of Information	Establishes Fair Information Practices: Right to See/Copy/Amend Own Record	Patient Consent for Disclosure is Informed and Voluntary	Few Exceptions to Informed Patient Consent	Restricts Creation of Cradle-to-Grave Medical Record	Strictly Limits Insurers' Access to Information
National CPR Opposes These Bills						
H.R. 1941 - Health Information Privacy Act, *Rep. Condit*	NO	YES				YES, but not strictly
H.R. 2470 - Medical Info. Protection and Research Enhancement Act, *Rep. Greenwood*	NO	YES	NO, legalizes widespread disclosure without patient consent for multiple purposes			NO, allows broad access
H.R. 2455 - Consumer Health & Research Technology (CHART) Protection Act, *Rep. Shays*	NO	YES				NO
S. 578 - Health Care Personal Info. Non-Disclosure Act, *Sen. Jeffords*	NO, facilitates widespread disclosure for multiple purposes	YES	NO, consent is coerced for care, payment, health-care operations	NO, many exceptions	NO	NO, breadth of purposes is quite large
S. 881 - Medical Info. Protection Act, *Sen. Bennett*	NO, facilitates widespread sharing	YES	NO, coerced consent unless patients pay for their own care			NO, insurers may use the info. broadly
U.S. Senate, Health, Education, Labor, and Pensions Committee (HELP Committee)	NO	YES				NO, insurers may use the info. quite broadly
Proposed U.S. Health and Human Services Department Regulations, *Executive Branch* (Applies to electronic records only)	NO	YES	NO, see *Condit*			YES, but not strictly; Establishes principle of disclosing "minimum info necessary"
National CPR Supports These Bills						
S. 573/H.R. 1057 - Medical Info. Privacy and Security Act, *Sen. Leahy, Rep. Markey*	YES, takes important steps to restrict widespread use of info.	YES	YES, consent is informed but not voluntary for treatment and payment. Self-pay option.	YES	NO	YES, but not strictly

Reprinted from National CPR Newsletter

	Establishes Higher Protection for Sensitive Information	No Preemption of Stronger Federal, State or Common Laws or Regulations	Requires Technical Security for Electronic Records	Restricts Employers' Access to Workers' Medical Records	No Routine Public Health Surveillance of Identifiable Medical Information	Research: Obtain Current or Prospective Patient Consent, Oversight by Medical Records Review Board	Stiff Civil and Criminal Penalties for Privacy Violations
National CPR Opposes These Bills							
H.R. 1941 - Health Information Privacy Act, *Rep. Condit*	YES, but allows HHS to set standards	YES	YES	YES, for employment-related decisions			YES
H.R. 2470 - Medical Info. Protection and Research Enhancement Act, *Rep. Greenwood*	NO	NO, preempts state law	YES	NO			YES
H.R. 2455 - Consumer Health & Research Technology (CHART) Protection Act, *Rep. Shays*	NO, see above		YES, but standards to be determined	NO		NO, broad access to records without patient consent	YES, but standards to be determined
S. 578 - Health Care Personal Info. Non-Disclosure Act, *Sen. Jeffords*	NO	YES, current laws stand — NO, states prevented from passing stronger laws after 18 months	YES				YES
S. 881 - Medical Info. Protection Act, *Sen. Bennett*	NO	NO, but allows existing state protections for mental and public health	YES	NO, not if self-insured	NO, allows broad use of patient information for routine surveillance		YES, but harmed individuals cannot sue
U.S. Senate, Health, Education, Labor, and Pensions Committee (HELP Committee)	NO	NO, but would preserve existing laws that are stronger	YES			NO, broad access	YES
Proposed U.S. Health and Human Services Department Regulations, *Executive Branch* (Applies to electronic records only)	NO, except psychotherapy	YES, does not preempt	YES	YES, for employment-related decisions		extends oversight	YES, and recommends patient right to sue
National CPR Supports These Bills							
S. 573/H.R. 1057 - Medical Info. Privacy and Security Act, *Sen. Leahy, Rep. Markey*	YES	YES	YES	NO	Creates some incentives to use de-identified info.	NO, broad access but extends oversight	YES

83

in a coma could receive immediate lifesaving treatment because the medical record could be available at the click of a mouse.

On its face, the life/death argument appears indisputable. But it is unclear whether a narrow subset of emergency situations justifies having a medical record that is open to any clinician who wants to view it. Such an open record would preclude unbiased second opinions, anonymous testing (i.e., genetic) and a truly confidential treatment relationship. As discussed previously, the compromise would be significant.

The Massachusetts Medical Society's Special Committee on Information Technology recognized the potential privacy threat posed by electronic medical records and "large data repositories."[28] In early 1996, when the reference committee reviewed the resolution to develop a policy, it noted:

> "Because physicians have a direct responsibility to protect patient information, the Massachusetts Medical Society has a corresponding responsibility to take an active role in formulating policy, as well as educating physicians and non-physicians about patient confidentiality in the information age."[29]

A Presidential Task Force on Privacy and Confidentiality was formed, and in November 1996, the Massachusetts Medical Society adopted a landmark policy containing several key general principles:

- The patient has a fundamental right to privacy and confidentiality in his/her relationship with a physician.
- Privacy and confidentiality are the patient's privilege, to be waived only by the patient, in a meaningful, non-coerced way.
- Conflicts between a patient's privacy right and a third party's need to know should be resolved in the patient's favor unless that would result in serious harm to the patient or others.

- Statutory exceptions to privacy/confidentiality should be narrowly defined.
- The development of new information technologies should include measures that strengthen patient privacy and confidentiality.
- Physicians have an ethical duty to understand these issues, educate staffs, and make reasonable efforts to keep patients informed.

In addition, the Society made strong patient-centered recommendations about the use of electronic medical records for clinical care:

- The benefits of information technology for care and research will not be realized without strong policy to guide its use.
- Access to the clinical record for care should be obtained through the attending physician according to the patient's consent.
- Access to patient data should be controlled by security features and tracked to a readily available audit trail.
- Patients have the right to determine (in consultation with their physicians) what data should/should not be disseminated electronically.
- Systems that encourage data linkage (i.e., unique health identifiers) may jeopardize patient privacy and should require the patient's consent.
- Firm statutes should regulate access to electronic patient data and establish penalties for security violations.[30]

In 1998, the American Medical Association followed suit and adopted a "vital new policy to protect patient privacy."[31] Like the Massachusetts Medical Society, the AMA upheld a patient's right to privacy, to be waived only by the patient or in "rare instances of strongly countervailing public interest."[32] The AMA also opposed the mandatory assignment and use of unique patient identifiers because of their potential threat to patient privacy.[33]

Other professional organizations, such as the American Psychiatric Association, the American Psychoanalytic Association, and the American Association of Physicians and Surgeons have also taken similar positions in the medical privacy debate. In its preliminary comments about HHS' proposed regulations, the American Psychoanalytic Association reaffirmed that individuals have a right to medical privacy, which should not be waived without meaningful notice and informed consent.[34]

In addition, the Surgeon General, in his December 1999 "Report on Mental Health," made a compelling case for strongly protecting the confidentiality so vital to successful mental-health treatment. Quoting the US Supreme Court decision, *Jaffe v. Redmond*, the report said:

> "The psychotherapist privilege serves the public interest by facilitating the provision of appropriate treatment for individuals suffering the effects of a mental or emotional problem. The mental health of our citizenry, no less than its physical health, is a public good of transcendent importance."[35]

The conviction that confidentiality was a necessary condition for any quality treatment (not just psychiatric care) prompted several concerned individuals to establish the National Coalition for Patient Rights (National CPR). Founded in 1994 because of the erosion in patient privacy, National CPR has been dedicated to restoring confidentiality to health care. The organization has worked to "raise public awareness through advocacy and public education, prevent infringements on privacy" [36] and pass truly protective legislation. At the outset, its goals were based on five basic principles:

- Requiring informed consent from patients for any information to be entered into or accessed from a computer network.

- Protecting patients and providers from penalties for declining to computerize medical information.
- Allowing patients to receive medical care and payment without using a national ID number.
- Strictly limiting the amount of clinical information required by insurers.
- Providing a mechanism for confidential, independent clinical review if insurers dispute medical necessity.

The recommendations put forth in the recent National CPR White Paper, "Protecting the Privacy of Medical Records: An Ethical Analysis" represent the development of patient-centered, ethics-based principles into sound, practical policy (*see* Fig. 2).

The major premise of this ethical analysis and policy statement is that the medical record is primarily a tool to document and facilitate patient care. Patients disclose information to their caregivers because they believe it will be used for their clinical benefit. The utilitarian rationale for keeping health information private and confidential is that this allows for the trusting, open communication necessary for an accurate, timely clinical history. Because tainted or incomplete data would undermine the information on which care or research is based, privacy and confidentiality are cornerstones of quality care. The duty-based argument for medical privacy lies in patients' expectation that their caregivers will respect their autonomy, control, and dignity. There is an ethical obligation inherent in this, which forms the basis for the time-honored practice of informed consent. [37]

Based on these principles, the White Paper authors recommended that "medical records should be maintained as confidential and private for the purpose of the clinical benefits of the patient. Disclosure.... outside the context of clinical care requires consent of the patient."[38] For privacy in the clinical care context, the following recommendations were made:

Recommendation 1:Medical records should be maintained as confidential and private for the purpose of the clinical benefits of the patient. Disclosure of medical records outside the context of clinical care requires the consent of the patient.

Recommendation 2: The right of patients to determine what information in their medical records is shared with other providers and other institutions and agencies should be recognized both by law and by institutional policy. Patients who wish not to disclose medical information to other health care providers that may be important in their medical care should be counseled about the risks of nondisclosure and sign an acknowledgement of their being warned.

Recommendation 3: Patient's should have the legal right to review and copy their medical records, as well as correct and amend errors. Patients should have the right to review a disclosure record.

Recommendation 4: Third party payers of medical services should be required to specify in advance the medical information they require to assess claims and manage medical care. Physician notes should not routinely be disclosed to third party payers. Disease management programs should be based on sound clinical research and arranged, with the patient's consent, through his/her health care provider.

Recommendation 5: Third party payers should be held accountable to the same standards of privacy and confidentiality as are medical care providers. No re-disclosure may be made without the written *freely given consent* of the patient, and participation in the health plan or other benefits should not be contingent upon patient consent to further disclosures.

Recommendation 6:The psychotherapeutic relationship is of such sensitivity as to require special recognition as a domain of absolute privacy. Consistent with Jaffe v. Redmond, records and notes of psychotherapy sessions should always remain confidential and third parties should be prohibited by law from demanding their disclosure for any reason. The minimum amount of information necessary should be disclosed to process claims.

Recommendation 7: Research involving medical records must either be conducted with the freely given informed consent of patients, or with blanket consent which delegates to a Medical Records Review Board (MRRB) the authority to waive further consent. The MRRB should be constituted by at least a majority of community members in addition to appropriate scientific, medical and allied health personnel and administered by the Medical Records Trustee. MRRB decisions not to grant a waiver of informed consent should be final. The MRRB should insure that the confidentiality of patient information is protected as it passes through a research protocol, that the information is not used for other purposes without explicit MRRB approval, and that the purposes of research will not be reasonably objectionable to the patient populations involved.

Figure 2 *(continued on opposite page)*

- The right of patients to determine what information in their medical records may be shared with other providers and other institutions and agencies should be recognized both by law and by institutional policy. Patients who wish not to disclose medical information to other health-care providers that may be important in their medical care should be counseled about the risks of nondisclosure and sign an acknowledgment of their being warned.

Recommendation 8: All health services research that relies on personal medical information should be reviewed, approved, and overseen by an institutional Medical Records Review Board, as with biomedical research.

Recommendation 9: Each clinical institution maintaining medical records has the responsibility to safeguard their confidentiality by minimizing access to medical records to those individuals whose "need to know" is of clinical benefit to the patient or is otherwise consented to by the patient. Institutions should employ encryption schemes, password protection, and computerized audit trails. Institutions should routinely audit to ensure that access to medical records is appropriate and take punitive measures when it is not. Patients should have the right to limit access to particularly sensitive information.

Recommendation 10: Each health care institution maintaining medical records or medical information should designate a "Medical Records Trustee" responsible for promulgating and enforcing institutional confidentiality and privacy policies, and ensuring compliance with the law.

Recommendation 11: Public health investigations in which an imminent danger to the health of individuals or communities is at stake, should be permitted to access private medical records as necessary and as provided for under current law. When providers make legally mandated disclosures to public health authorities they inform the patient of this requirement when the condition is discovered. Surveillance and epidemiological studies should be guided by the recommendations for biomedical research. (Recommendation 7)

Recommendation 12: In general, employers should not have access to clinical medical records. These records should be segregated from all other personnel-related information, and be used only in the benefits determination process (only if employer is a self-insurer). Employers should be barred from using this information for employment-related decisions.

Recommendation 13: Health care institutions maintaining medical records should notify the public and patients individually of the offices and functions which have access to their medical records. Institutions should also prominently display their policies on maintaining confidentiality of medical records, including the name, address, and phone number of the Medical Records Trustee.

Recommendation 14: Proposals to create systems designed to link private medical information or otherwise collate medical record information, such as the Unique Patient Identifier or the Master Patient Index, should not be implemented without explicit patient informed consent.

Recommendation 15: Law enforcement access to medical records should be limited to court order. When records are thus obtained, they should contain only the minimal amount of information necessary to fulfill the purpose for which they were sought. In health care fraud investigations, anonymous records should be used to assess patterns of fraudulent billing, with identified information used only where specific instances of fraud are suspected.

Recommendation 16: The buying and selling of medical records or information derived from them, and the use of these records for any marketing purposes, including disease management programs, without the freely given informed consent of the patient, should be prohibited.

Fig. 2. *(continued)*

- Each clinical institution maintaining medical records has the responsibility to safeguard their confidentiality by minimizing access to medical records to those individuals whose "need to know" is of clinical benefit to the patient or is otherwise consented to by the patient. Institutions should employ encryption schemes, password protections, and computerized audit trails. Institutions should routinely audit to ensure that access to medical records is appropriate, and take punitive measures when it is not. Patients should have the right to limit access to particularly sensitive information.

- Each health-care institution maintaining medical records or medical information should designate a "Medical Records Trustee" responsible for promulgating and enforcing institutional confidentiality and privacy policies, and ensuring compliance with the law.[39]

These patient-centered policies take a clear position that medical records ought not to be broadly available even for clinical care without the explicit knowledge and consent of the patient. Exceptions to this should be strictly and narrowly defined, with the patient almost always retaining the right to limit disclosures.

In a Health-Care System Largely Financed by Third Parties, How Can Policy Guide Their Access to Patient Information While Preserving a Confidential Clinical Relationship?

The Massachusetts Medical Society's position on payer access was resolutely in favor of patient privacy and limited insurance-company access, as follows:

- Physicians' participation in and patients' enrollment in a health plan should not compel a blanket, broad release of health information.
- Third-party payers should limit the scope of information required to the minimum necessary for the specific function, and should not routinely seek the entire medical record.

- The patient's written consent should be obtained each time a third-party payer seeks access to an individual's medical information.

Similarly, the AMA bolstered the principle of the "minimum information" to be disclosed to a third party for the specific purpose requested. Strict guidelines were set (similar to the Massachusetts policy) for insurers' and employers' access to identifiable patient information.[40] Several states have statutes that strictly limit the disclosure of mental-health information to payers.[41] The Surgeon General's Report on Mental Health suggests that these laws may be a model for others to follow as they develop policy in this area.[42]

In its White Paper, National CPR expanded on earlier principles and organized medicine's policies about third-party payers, as follows:

- Third-party payers of medical services should be required to specify in advance the medical information they require to assess claims and manage medical care. Physician notes should not routinely be disclosed to third-party payers. Disease management programs should be based on sound clinical research and arranged, with the patient's consent, through his/her health-care provider.
- Third-party payers should be held accountable to the same standards of privacy and confidentiality as are medical-care providers. No re-disclosure may be made without the written *freely given consent* of the patient, and participation in the health plan or other benefits should not be contingent upon patient consent to further disclosures.
- The psychotherapeutic relationship is of such sensitivity as to require special recognition as a domain of absolute privacy. Consistent with *Jaffe v. Redmond*, records and notes of psychotherapy sessions should always remain

confidential and third parties should be prohibited by law from demanding their disclosure for any reason. The minimum amount of information necessary should be disclosed to process claims.[43]

Clearly, limited access to third-party payers is unanimously endorsed in these privacy policies, while creating incentives for accountability and security for these entities. The sensitive nature of psychotherapy communications leads to a higher level of protection for these records, consistent with *Jaffe v. Redmond* and The Surgeon General's Report on Mental Health.

Can We Ensure the Continuation of Valuable, Needed Research and Public Health Initiatives Without Trampling on People's Rights?

In recent years, numerous factors have contributed to a strong push to weaken earlier ethical guidelines for research with human subjects. Most biomedical research with humans has been accountable to ethical standards contained in the Nuremburg Code (1947), the Declaration of Helsinki (1964)[44] and the Belmont Report (of the National Commission for the Protection of Human Subjects in Research, 1978).[45] The first two documents "are alike in their insistence that patient autonomy be respected and supported and in their elevation of concern for the rights of the individual patients and research subjects above scientific and societal goals."[46] Similarly, the Belmont Report recommended that "the rights of subjects were to be respected, specifically by obtaining their informed consent to participate."[47] The National Commission was also sufficiently concerned about the implications of using medical records for research, such that they defined that practice as research with human subjects. They recommended that the Department of Health Education and Welfare (HEW) require full Institutional Review Board review of all research using medical records.[48] Unfortunately, HEW, and subsequently Health and Human Services (HHS) did not follow that recommendation, and medical records are increasingly sought and obtained without people's knowledge or consent for research.

In the meantime, the research community is forging ahead to weaken existing guidelines for informed consent and IRB waivers for consent. As a result of the Office for the Protection of Research Risks' (OPRR) push to expand the definition of "minimal risk," research using medical records (and existing tissue samples) could be exempted from informed consent requirements as well as a full IRB review.[49] IRB workloads and financial incentives prime the pump for expedited reviews, and well-intended researchers worry that stringent consent requirements could irrevocably thwart and bias biomedical research.[50]

While it is crucial to acknowledge that there may be a price to pay for protecting privacy in medical record research, the ethical and clinical implications of failing to do so are even more compelling. The proposed elimination of patient control over the use of their records for research will effectively dismantle a fundamental ethical principle required of researchers: respect for the subject's dignity, autonomy, and control (as seen in the Nuremburg Code, Declaration of Helsinki, and the Belmont Report of the National Commission). This will threaten the integrity of the information used for care and research by undermining patients' trust in the system. It also sets the stage for a dangerous downhill slide in research ethics in the United States, with potentially disastrous consequences.[51]

The Massachusetts Medical Society recognized this in their recommendations for medical-record research:

- Medical information used for research should be de-identified at the source unless the patient consents for it to be disclosed/used for research.
- Information used for research must be protected from redisclosure without the patient's consent.

- For public health initiatives, information should be collected on a disease- or condition-specific basis only, and not for routine public-health surveillance.[52]

The AMA suggested in their policy that whenever possible, de-identified information should be used for research.[53]

National CPR's White Paper built on the Medical Society policy as well as the original National Commission recommendation for full IRB review of projects, in an effort to respect privacy and dignity while advancing needed research. Rather than requiring specific patient consent whenever records are sought, the authors suggested an alternative: upon entering a health care facility or enrolling in a health plan, patients would have the option to sign a prospective blanket consent for their medical records to be used for research. This authorization would be delegated to a Medical Records Review Board, a body with a majority of community representation, to review each proposal for its merits and privacy implications. This system would relieve already overworked IRB's of a large additional burden, and would provide a practical, respectful way to allow patients control of their personal information. The authors applied this model to health services research as well as public health surveillance and epidemiological research.[54] Furthermore, in cases of imminent danger to the health of individuals or communities, public-health investigators should be able to obtain access to private medical as provided for under current law. When providers make these legally mandated public health disclosures they should inform the patient of this requirement at the time the condition is discovered."[55]

In addition to the specific policy recommendations made by the Massachusetts Medical Society, AMA, American Psychiatric Association, American Psychoanalytic Association, and National CPR, these groups uniformly favor federal legislation that would *not* preempt stronger state privacy laws. This would provide a federal floor of protection, and allow states to enforce or enact stronger protections for their own citizens. The mandatory implementation and use of a national-health identifier or other data-linkage system is another area for concern because "such systems violate the liberty of patients to each determine for themselves who shall see their medical records and for what purposes."[56] Therefore, proposals to create such systems "should not be implemented without explicit informed consent."[57]

Conclusion

The ethical dilemmas presented by the Information Age are clear, but not insurmountable. Growing evidence supports a finding that compromised clinician–patient confidentiality is bad for both the individual and society's health. Short of a few, narrowly defined compelling reasons (i.e., imminent harm to oneself or another, and statutory reporting requirements), there is little justification to override patient privacy. We must therefore develop sound, patient-centered policies about health information, based on the ethical duty to respect the individual's autonomy, dignity, and control over health-care decisions.

Some core components of strong, patient-centered policy include a federal floor of protection, informed non-coerced patient consent for disclosures of health information (with few exceptions), the right to decline use of a unique patient identifier or linked electronic record system, and policies to limit the information disclosed to payers. Uses of information for laudable purposes, such as research and public-health surveillance, should be guided by either patient consent or a suggested model of prospective patient consent delegated to an appropriate review board. Most important, the right of patients to privacy in their relationship with their health care professional needs to be explicitly acknowledged and legally protected. Several model policy proposals are presented, with the hope that public policy moves in that direction.

Acknowledgment

The author wishes to acknowledge Kathleen Osborne Clute for her tremendous help in preparing this manuscript.

References

[1]Glaser, J. (1996) "Patient Confidentiality: Enhancing the Power and Privacy of Medical Information." Harvard Pilgrim HealthCare conference, Boston, MA, September.

[2]A Guide to State-Level Ambulatory Care Data Collection Activities (1996) National Association of Health Data Organizations, Falls Church, VA.

[3]O'Harrow, R. (1998) Prescription sales, privacy fears. *The Washington Post*, February 15, p. A1.

[4]National Coalition for Patient Rights (CPR) Website. Confidence Betrayed. http://www.nationalcpr.org

[5]National Coalition for Patient Rights (CPR), Winter 1999 newsletter. Available at http://www.nationalcpr.org

[6]Goldman, M. (1999) Personal communication.

[7]Nagel, D. (1999) Personal communication.

[8]Linowes, D. F. and Spencer, R. C. (1998) How employers handle employee's personal information: report of a recent survey. *Employ. Rights Employ. Policy J.* **1(1),** 153–172.

[9]Cole, C. (1999) High rates of asthma inspires new study. *The Boston Globe*, Northwest weekly section, October 24.

[10]Dick, R. and Steen, E. (eds.) (1991) *The Computer-Based Patient Record: An Essential Technology for Health Care.* National Academy of Sciences, Washington, DC.

[11]National Research Council (1997) *For the Record: Protecting Electronic Health Information.* National Academy of Sciences, Washington, DC, p. 1.

[12]California HealthCare Foundation (1999) Medical privacy and confidentiality survey. Summary and overview, January 28: http://www.chhf.org/conference/survey/cfm

[13]Kremer, T. G. and Gesten, E. G. (1998) Confidentiality limits of managed care and client's willingness to self-disclose. *Prof. Psychol. Res. Prac.* **29(6),** 553–558.

[14]Stoekle, J. (1996) *Policy on Patient Confidentiality and Privacy.* Massachusetts Medical Society, November 8, p. 10.

[15]Jaffe v. Redmond, US Supreme Court, 1996.

[16]Ford, C., Millstein, S., Halpern-Feisher, B., and Irwin, C. (1997) *JAMA* **278(12),** 1029–1034.

[17]Naser, C. R. and Alpert, S. A. (1999) Protecting the privacy of medical records: an ethical analysis. A White Paper commissioned by The National Coalition for Patient Rights, Lexington, MA, May.

[18]Ibid.

[19]Alpert, S. (1998) Health care information: access, confidentiality and good practice, in *Ethics, Computing, and Medicine* (Goodman, K. W., ed.), Cambridge University Press, Cambridge, UK, p. 75–101.

[20]Wofford-Dodd Amendment (Administrative simplification) and Condit bill (Fair information practices), 1994.

[21]Kolata, G. (1995) When patient records are a commodity for sale. *The New York Times*, November 15, p. 1.

[22]Health Insurance Portability and Accountability Act of 1996, Public Law 104–191, enacted August 21, 1996, Washington, DC.

[23]Conference brochure. American Health Informatics Association, Computer-Based Patient Record Institute, Joint Health-Care Technology Alliance (1997) (out of print).

[24]Bills texts accessible through http://www.nationalcpr.org

[25]National CPR Winter 1999 newsletter.

[26]Billing code: 4150-04MD Department of Health and Human Services. Office of the Secretary. 45 CFR Parts 160 through 164. Rin: 0091-AB08. http://aspe.os.dhhs.gov/admnsimp.

[27]National CPR Winter 1999 newsletter.

[28]Policy on Patient Confidentiality and Privacy (1996) Massachusetts Medical Society. November 8, p. 6 (Nagel, D., M.D. consultant to Task Force).

[29]Ibid., p. 2.

[30]Ibid., p. 2–4.

[31]National CPR Summer 1998 newsletter, p. 2.

[32]Ibid.

[33]Report of the Board of Trustees (1998) Patient Privacy and Confidentiality, presented by T. R. Reardon, MD, chair, to Reference Committee on Amendments to Constitution and Bylaws, September 1998, Chicago, IL. www.ama-assn.org/meetings/public/annual 98/repats/bot/bot09.htm.

[34]American Psychoanalytic Association (1999) Draft Comments on Standards of Privacy of Individually Identifiable Health Information, December 15, New York, NY.

[35]US Dept. of Health and Human Services, (1999) *Mental Health: A Report of the Surgeon General,* Chapter 7, "Confidentiality." US Department of Health and Human Services, Substance Abuse and Mental Health Services Administration, Center for Mental Health Services, National Institutes of Mental Health, National Institutes of Health, Rockville, MD.

[36]National CPR brochure, p. 2. Available by request at http://www.nationalcpr.org

[37]Naser, C. R. and Alpert, S. A. (1999) Protecting the privacy of medical records: an ethical analysis.

[38]Ibid., p. 56.

[39]Ibid., p. 56.

[40]Report of the Board of Trustees (1998) Patient Privacy and Confidentiality.

[41]Massachusetts Annotated Laws, Chapter 8 Revision (1996) and New Jersey Statute NJSA (1985) 45:14B-31 et seq.

[42]The Surgeon General's Report on Mental Health, p. 449.

[43]Naser, C. R. and Alpert, S. A. (1999) Protecting the privacy of medical records: an ethical analysis.

[44]Woodward, B. (1999) Challenges to human subject protections in US medical research. *JAMA* **282(20),** 1947–1952.

[45]Brennan, T. A. (1999) Proposed revisions in the Declaration of Helsinki: will they weaken the ethical principles underlying human research. *NEJM* **341(7),** 527–530.

[46]Woodward, B., p. 1947.

[47]Brennan, T. A., p. 528.

[48]Naser, C. R. and Alpert, S. A. (1999) Protecting the privacy of medical records: an ethical analysis.

[49]Woodward, B. (1999) Personal communication.

[50]Melton, L. J. (1997) The threat to medical records research. *NEJM* **337(20),** 1466–1470.

[51]Goldman, M. (1999) Presentation to Department of Energy's Human Subject Research Group, Boston, MA, December.

[52]Policy on Patient Confidentiality and Privacy (1996) Massachusetts Medical Society. November 8, p. 4 (Nagel, D., M.D. consultant to Task Force).

[53]Report of the Board of Trustees. (1998) Patient Privacy and Confidentiality.

[54]Naser, C. R. and Alpert, S. A. (1999) Protecting the privacy of medical records: an ethical analysis.

[55]Ibid.

[56]Ibid., p. 63.

[57]Ibid., p. 63.

Abstract

We live in an era in which the growth of powerful information technology; the profound changes in our system of health-care organization, delivery, and financing; and dramatic advancements in biomedical research, especially in human genetics and genomics, have stirred intense public concern about the privacy of individually identifiable medical information. Protection of medical-information privacy in the United States at this time is limited to State laws, which are generally regarded as fragmentary, variably effective, discordant, and dissatisfying. Why, despite broad-based public support of the need for comprehensive federal legislation, have repeated Congressional efforts in recent years to enact privacy legislation failed to get out of committee? This chapter attempts to answer this question by providing context that is often lacking in the public and political discourse. I explain the high volume and complex flows of identifiable medical information that are necessary for the day to day operations of the contemporary health-care system, that is, to provide treatment and payment and enable system operations, but give special attention to the critically important role played by archived medical information in supporting medical research, advancing medical knowledge, and improving the health of the public. Medicine is an empirical discipline, and the foundations of medical knowledge have been created over generations by the systematic study of countless patients and their families to elucidate the causes, natural history, presentations, and therapeutic responses of human diseases. I emphasize that it is possible to provide identifiable medical information used in research with almost total protection from forcible trespass, and indeed, to assure the confidenti-

ality of such information with far more confidence than is true for medical information used in providing care, or, for that matter, for any other kind of personal information. I argue that everyone who confronts disease is an immediate and direct beneficiary of the knowledge accrued over generations from studies using medical information of countless other individuals and families with similar ailments, and that the ethical principle of justice supports the proposition that since everyone benefits, everyone is obligated to allow his or her medical information to be used in medical research.

Medical Information Privacy and the Conduct of Biomedical Research

David Korn

Introduction

We are living in a remarkable era, in which a number of profound, and often unsettling, changes are taking place, all of which interact and bear on medical information privacy, the practice of medicine, and the conduct of medical research. In using the term "medical research," I mean ecumenically to embrace the entire gamut of what the recent report of the Association of American Medical Colleges (AAMC) Task Force on Clinical Research has proposed as a new and inclusive (and, I might add, hospitable) definition of clinical research, namely, studies ranging from the etiopathogenesis of disease, translation, and clinical trials, to research in epidemiology, prevention, and health services, all of which are critically dependent on access to medical and health information.

What Are Some of These Changes?

First, we live in an "age of knowledge" characterized by the remarkable development and permeation into all sectors and lay-

ers of society of electronic information technology that is trans-
forming the global economy and creating new industries and vast
new wealth based almost entirely on the capability of collecting,
manipulating and strategically using personal information. The
technology relentlessly increases in power and decreases in cost,
and has enabled the creation of huge databases that are trans-
missable, linkable, and relatively easily accessible. These databases are repositories of very personal, very private information
about all of us and include such tasty morsels as bank records,
credit card records, Department of Motor Vehicles records, all
sorts of legal records, telephone records, and on and on, seemingly without end. Recognition of this fact has created a pervasive sense of unease, and anger fueled by an equally pervading
sense of helplessness, about the "loss of individual privacy."
Evidence of this unease is all about us, for example, in a recent
report (1998) on the topic from the Center for Public Integrity that
was entitled simply, "Nothing is Sacred;" or in the vivid (perhaps
a better word is lurid) reports in recent years on the front pages
of the *New York Times*, the *Washington Post*, and other major
newspapers, as well as news magazines. Simply put, these various articles have made abundantly clear that by means both licit
and illicit, and for relatively small amounts of money (although
illicit costs somewhat more), any one of us can find out just about
anything we wish to know about just about anyone else, including
gaining access to their medical records.

     Second, we live in an era in which massive changes are
occurring in our system of health-care delivery and financing,
changes that I refer to as the "industrialization" of medicine.
These changes are characterized by an intense societal (but, an
important disjunction, not individual) focus on health care costs,
consolidations of purchasers, insurers, and providers, and heavy
investments in information technology. The latter have permitted
the creation of very large patient databases that can be, and are,
transmitted easily and widely, and can be, and are, readily accessed
by many parties for a host of purposes relating to the delivery of

care, the payment for care, and the operations of the health-care delivery system itself. Most of these purposes are legitimate and arguably necessary, but others appear to many to be less so and challengeable. For purposes of this chapter, it is important to understand that the terms "payment" and "operations" are remarkably flexible and capacious. They embrace an array of activities, functions, and personnel, all of which claim a compelling need to access identifiable medical records, a fact that the average citizen does not understand and finds alarming.

Third, we live in an age of truly astonishing advancements in biomedical knowledge and technology, especially in genetics, that are providing powerful new capabilities and insights into human health and disease. As but one example, consider the rapidly unfolding saga of the intricate genetic pathways that underlie human neoplasia, as exemplifed by the delineation of the genetic progression of colorectal cancer, largely work from the laboratory of Professor Bert Vogelstein at Johns Hopkins,[1] or the recent tour de force published by a group of Stanford investigators that profiles gene expression in diffuse large B-cell lymphoma.[2,3] Progress in cancer genetics has been so rapid that the National Cancer Institute several years ago was able to launch a major new initiative, the Cancer Genome Anatomy Project, with the goal of cataloguing all of the genetic changes that occur in human neoplasms. As recently as 1991, my last year as chair (1984–91) of the National Cancer Advisory Board, even to have suggested such an ambitious, indeed, audacious, undertaking would have been thought delusional.

Advances in genetics and genomics are creating data at an astounding rate and have stimulated the development of the entirely new discipline of bioinformatics, which has the potential to transform not only the conduct of basic science but the understanding and practice of medicine as well. These advances have also stirred powerful new fears within the public and poured high-octane fuel on the already widespread and smoldering concerns about the loss of individual privacy. They have also spurred concerns in both federal and state legislatures, and have led to enactment in more than half the states of hastily crafted, often emotionally driven, and poorly conceived legislation that, in highly variable ways, and with equally variable mixtures of success and unintended consequences, attempts to limit access to and use of genetic information.

The combination of these changes has spurred a crescendo of public anxiety that begins with unease and frustration about the perception of a steadily progressive erosion of privacy in general, accentuates dramatically when it comes to medical-information privacy, and goes off the scale at the mention of "genetics," a term that an average citizen might poorly understand and generally fear.

While wrestling with and debating these very difficult and emotionally charged issues over the past several years, I have come to recognize two quite different roots of public concern: one, I shall call "pragmatic," which is concern about such things as loss of health insurance, discrimination in employment, and social stigmatization. The second root is "ideological" and springs from a strong, deeply held belief in an individual's right to privacy, from which arise such positions as an individual's right to have complete control of his/her medical information, and the proposition advanced by the respected legal-ethicist, Professor George Annas of Boston University,[4] that individuals have property rights in their genetic information and, therefore, in any biological materials, samples, and tissue scraps that in today's state of science and technology have become "genetic repositories."

These two different roots, one pragmatic and the other ideological, require, in my view, two quite different types of responses. The former can, at least in theory, be dealt with effectively by the political process with appropriate statutes and regulations that circumscribe the uses of personal medical information and prohibit discrimination on the basis of particular components of it, for example, predictive genetic tests. Thus, in February 2000, President Clinton issued an Executive Order prohibiting the dis-

criminatory use of genetic tests or information in federal employment. Of interest is the fact that this Order has elicited mixed
reaction from commentators: much of the criticism has centered
(not surprisingly) on the operational ambiguity of the term "genetic
information" and fears of the unintended overreaching that can
result.

Unfortunately, ideologies of whatever sort tend to fuel
extreme positions that make rational discourse much more difficult and confound attempts at political compromise. It is important to understand that the privacy ideology exists and is deeply
held, and that it is capable of distracting and confounding public
discourse and political debate across the spectrum of biomedical
research and medical practice, with an impact quite analogous to
that of "right-to-life" and "animal rights" doctrines. To this point,
let me make clear that I strongly support the statement made by
DHHS Secretary Donna Shalala in September 1997 in her report
to the Congress on medical information privacy, that in contemporary society there is and can be no absolute right to privacy.

Medicine Is an Empirical Discipline

Medicine is, and will remain, fundamentally an empirical
discipline. Much of the contemporary base of medical knowledge has been constructed over the generations from the systematic study of individuals and collections of individuals clustered
by symptoms, signs, and ultimately, diagnoses, which progres-
sively become more refined as etiopathogenic mechanisms are
illuminated, the biological roots of human disorders more firmly
established, and nosology more precise and informative. The
systematic study of diseased patients is an imperative deeply
embedded in the ethos of medicine, and it goes back centuries,
probably to the 14th century, when Italian physicians began to
conduct autopsies to try to glean understandings from their
deceased patients that might better equip them to deal with simi-

lar patients yet to be encountered. The introduction of autopsies at that time probably represented the very first glimmerings of the dawn of scientific medicine.

These kinds of systematic studies, clinical and pathological, which have provided us with all of the vocabulary and most of our knowledge about the definition, causation, natural history, and therapeutic responsiveness of human diseases, have required—indeed, been absolutely dependent on—access to patient records and often, stored tissue samples obtained during the course of delivering medical care. The materials, commonly dubbed "archival patient materials," have throughout the history of medicine been considered a *public* archive—a *public* research resource—that is unique, invaluable, irreplaceable, and in a sense, immortal, and constitutes a vast research library of medical experience over the generations.

Today, in the new era of genetic and molecular medicine, information technology, and managed care, with increasing emphasis on disease prevention, population health, and the elimination of health disparities, with increasing demand for the practice of "evidence-based medicine," and facing the monumental task of phenotyping the thousands of putative disease-associated genes that are being revealed by the Human Genome Project, these archival patient materials become even more valuable as the unique research resource that provides the essential foundation for most, if not all, of the studies that will be necessary to accomplish these lofty objectives.

Federal Oversight of Research Involving Human Subjects

In the aftermath of World War II and the Nuremberg trials, all Western nations have developed statutes and regulations aimed at protecting human subjects who participate in research. These protections in the United States were initially codified in 1974 in the Code of Federal Regulations (45CRF46) by the Department of HEW (now HHS) and were later adopted, in 1991, by all 17 federal agencies that conduct or sponsor human subject research, at which time they became known as the "Common Rule" (56FR28003). The Food and Drug Administration (FDA) is not bound by the Common Rule but has long operated under its own separate, but almost identical, set of regulations dealing with this issue (21CFR50). Thus, a comparable code of protections governs all research involving human subjects that is conducted or sponsored by the federal government, as well as any research conducted in academe or by the biotechnology, pharmaceutical, and medical-device companies that falls under the jurisdiction of the FDA. In academic medical centers, all research involving human subjects, no matter how funded, is typically covered by the Common Rule. Although compliance with the Common Rule is not perfect, the rule has earned an important degree of public trust, and the fact that in non-academic venues some kinds of human subject research may not be covered by the rule turns out to be a significant source of public concern that frequently arises in debates about medical information privacy (or the ethics of medical research more generally).

Several features of the Common Rule are especially pertinent to this essay. First, what is a "human subject?" The Common Rule states that "human subject" means a living individual about whom an investigator obtains data through intervention or interaction with the individual or identifiable private information, for example, a medical record. Second, the Common Rule requires that human research participants provide the investigator with informed consent (to which I shall return), and that all research on human subjects be reviewed by an Institutional Review Board (IRB). Third, the Common Rule permits informed consent to be waived if the research involves no more than minimal risk, if the waiver will not adversely affect the rights and welfare of the subjects, and if the research could not practically be carried out without the waiver. Fourth, the Common Rule permits abbreviated, or "expedited" review of several types of research, including that which involves "materials (data, documents, records, or specimens) that have been collected or will be collected solely for nonresearch purposes (such as medical treatment or diagnosis)." Fifth, among the specific criteria that an IRB review must include are the determinations that risks to subjects are reasonable in relation to anticipated benefits, if any, and to the importance of the knowledge that may reasonably be expected to result from the study; and that when appropriate, there are adequate provisions to protect the privacy of subjects and maintain the confidentiality of data.

Notwithstanding the provisions just summarized, I will argue that the major features of the Common Rule have been designed for prospective studies of human subjects, where it is possible for the investigator to interact with and "fully inform" prospective participants of the exact nature of the study, its duration, risks and possible benefits, matters of conflict of interest, and so forth. Indeed, the Common Rule enumerates 14 specific elements that must be covered in the informed consent process. The requirement of "full disclosure" to the prospective research participant lies at the very heart of the concept of informed consent, and, in turn, determination of the adequacy of informed consent and the assessment of risk by the IRB lie at the heart of the Common Rule.

I will also argue that historically, research that is retro–spective and requires access to archival patient materials but involves no interactions with patients themselves, which is variously described as secondary, non-interactional, or non-interventional research, has typically been considered by all participants in the process to be of no more than minimal risk, and has been conducted without notification of the IRB, or, more properly, under IRB waiver or expedited review. I will further assert that such research has in fact been of no more than minimal risk to any patient and of immense public benefit; and that every individual and family that confronts illness is the immediate and direct beneficiary of the vast body of medical knowledge that has been derived from research on other individuals' and families' experiences, and has been accumulating over the generations *pro bono publico*.

The ethical principles that underlie the conduct of research on human subjects are conveniently summarized as respect for persons (or autonomy), beneficence, and justice; the last is interpreted to mean that the benefits and burdens of research must be distributed fairly. I believe the argument can be fairly made that since every individual is a direct beneficiary of the historic medical knowledge base, the ethical principle of distributive justice would suggest that everyone should be obligated to contribute to the ongoing renewal of that base.

The Great Public Debate

Genetic Information

I turn now to the great public debate raging over medical information privacy, and the control of access to and use of such information. My baptism in these issues occurred in 1995, when I first became immersed in the debate about research access to human tissue samples. (I began my professional life as a pathologist, and I was enlisted into this battle by the American Society for Investigative Pathology, a member of the Federation of American Societies for Experimental Biology.) At that time, a number of professional groups and ad hoc committees, some initiated and/or supported by the Ethical, Legal and Social Implications (ELSI) Working Group of the National Human Genome Research Center (now Institute), had sponsored convocations or issued position papers on this topic, and the first stirrings of federal and state legislative interest in genetic privacy and discrimination were evident. The issues in contention concerned genetic testing and genetic research, the nature of the informed consent to be required, the kinds of uses and distributions of materials and test results, or research results, that should be permitted, and under what circumstances, and ultimately, who owned the materials and the data.

The core principles that seemed to be in favor at that time, and under which most of the participants seemed to be operating, included:

- Acceptance of the doctrine of what has been dubbed "genetic exceptionalism,"[5] that is, that genetic information, although seemingly very difficult to define operationally, is nonetheless unique and different from all other personal, private, sensitive and at times, predictive information that may be present in a medical record, and that it can and should be separately regulated and managed.

- More generally, information contained in a medical record can and should be segregated according to its perceived sensitivity and given separate security and rules of access. It was not always clear which information would be so judged, or by whom, other than perhaps by the advocates of the position. The kinds of information most often specified in these discussions included that relating to genetics, mental illness, and HIV positivity or sexually transmitted diseases (STDs) more generally, but, in reality, the list was unbounded. In my experience, individuals suffering from any major disease may consider their disease to be "unusually sensitive" and deserving of special protections. Indeed, Western history is replete with examples of disease-based discrimination, reaching back in the United States at least as far as the Salem witch trials and including in more recent times such disorders as epilepsy, mental disease, and cancer.

- Individuals owned their genetic information and the sources of that information (dubbed by one commentator,[4] "future diaries") and thus had near-absolute control of its accessibility, usage and fate, including its destruction.

The complexity, contentiousness, and seeming intractability of these issues led President Clinton to hand them as a first order of business to the newly created National Bioethics Advisory Commission (NBAC) at its first meeting in December, 1996. That body decided in its wisdom to parse the issues, and after more than two years of deliberation, in August 1999, published the first of its two-volume report on the subset of issues relating to the privacy of genetic information that concern research using stored human biological materials.[6] (The second volume consists of commissioned papers and includes one from me, entitled "Contribution of the Human Tissue Archive to the Advancement of Medical Knowledge and the Public Health"[7]). As has been true with earlier NBAC reports, it is not certain exactly what gravitas the latest issuance will have in shaping public policy, law, or regulation.

Because the public debate at the federal level in the past three years, and certainly the attention of the Congress, has shifted away from the protection of "genetic information," or any other special subcategories of medical information, to the broader, more encompassing, and certainly knottier issues of "medical information privacy," I shall put the matter of genetic information and access to human tissue samples aside with the comment that although federal attention to the issues of "genetic information" has somewhat subsided, public concern remains unabated; in no sense has this issue been fully or definitively resolved.

General Principles Espoused by the AAMC

In order to guide the thinking of the Association of American Medical Colleges about the very complex issues that are subsumed under the rubric of "medical information privacy," and to inform our educational and advocacy positions, my colleagues and I have articulated a set of eight overarching principles that have proved to be very helpful in dealing with the Congress and Administration, advocates of various stripes, and the general public.

1. The free flows of identifiable medical information within the boundaries of the health care delivery system are essential to the optimum provision of patient care and its payment, and for the operations of the health care system. Exactly where the boundaries of the system should be drawn is a matter of legitimate public concern and has been the subject of vigorous public discussion and political debate, that tend, in my view, to be most useful when informed by understanding of how the contemporary system of health care delivery and financing actually works. Knowledge, of course, has never been a precondition for expressing opinions in public policy or political debates. As a rule, the AAMC assesses with great caution any legislative or regulatory proposals that would impede these information flows or burden them with substantial new administrative complexities and costs.

2. Continued access to archival patient materials, properly overseen but not unduly hindered, is essential to the pursuit of the nation's medical and health research agenda. Accordingly, we oppose the erection of cumbersome and costly new barriers to research access to these materials. We take strong issue with the proposition that individuals have inherent property rights in their medical information or tissue samples, and the exceedingly restrictive, and in our view, socially maleficent, covenants that can flow from that construct.

3. There is an urgent need for federal legislation to establish a strong, comprehensive, and nationally uniform framework of protection of the confidentiality of identified or directly identifiable medical information, with stiff penalties for violations. In crafting such a law and its implementing regulations, it is essential that all involved bear constantly in mind that no statute or rule can have an efficacy of 100%, and that to reach for such an illusory goal will inevitably lead to devastating unintended conse-

quences. This seemingly self-evident truism is surprisingly hard to sell to some legislators and regulators.

4. There should be uniform protections for and rules of access to all medical records, without segregation of components perceived to be of "special sensitivity" for "special protections," with the exception of psychotherapy notes, the special status of which is well-established in medical practice and law. This stance tends to be controversial among some patient and disease advocacy groups that believe that information about "their disease," whatever it may be, is unusually sensitive and deserving of special protections. In my opinion, this issue is further confounded by the profound difference between how a lawyer regards a legal record and a physician or medical scientist thinks about a medical record. For the latter, the record is meant to be an accurate, candid, and complete picture of an individual's health and disease; for the former, the record is an artificial construct to be shaped and pruned in order to influence the decision of the trier of fact, who, judge or jury, is bound by rules of evidence to consider nothing beyond what is contained in the written ("admissable") legal record.[8]

    The AAMC especially opposes the concept of "genetic exceptionalism," which has not been laid to rest, as a matter of both principle and practicability: it is veritably impossible to operationalize this concept in contemporary medical practice or research; what exactly is a "genetic test," "genetic information," or "genetic research?" Or, perhaps better in the dawning age of human genetics and genomics, what is not?

5. Medical-information privacy law and regulation should be crafted purposefully to encourage the maximum possible use of "de-identified" medical information wherever possible, and especially in medical research. But to do so, the concept of "de-identification" and the defini-

tion of "de-identified" must be addressed with reason and sensitivity. In the current era of powerful information technology, it becomes increasingly difficult, if not impossible, to be certain of when a given set of medical information has been sufficiently "de-identified" to make the probability of its re-identification less than some very small predetermined value that would satisfy a legal criterion of "reasonableness." Experts in the field argue persuasively that the number and kinds of data elements contained in the typical medical record that would have to be deleted or distorted to satisfy such a criterion will only increase as computer power increases; and some argue further that in the near future, only substantially distorted health information might in fact prove to be unidentifiable. I would observe that the boundary between substantial distortion and fabrication can be tenuous. I would also note that excessive zeal in striving for some absolute level of "de-identification" will often so gut the resulting information as to render it useless for many kinds of record-based medical research.

For these reasons, and to ensure that "de-identified" medical information will remain maximally useful for research purposes, the AAMC argues that an appropriate definition of "de-identified" medical information would be information from which personal identifiers that *directly reveal*, or that provide a *direct means of identifying*, an individual have been removed or encrypted, such that the identity of an individual is not evident without use of an encryption key. Most medical research involving archival patient materials requires that the materials be linkable, both geographically and temporally. Therefore, we believe that it is very important for any new law, or rule, to recognize explicitly that encrypted medical information that remains linkable should be deemed "de-identified" or "non-identifiable" for purposes of statutory and regulatory compliance.

6. The AAMC strongly believes that Common Rule protections (or their equivalent) should be extended to all human subjects participating in research, regardless of the source of funding or the venue of that research. Furthermore, we argue that legislation or regulation addressing medical information privacy should, on first principles, defer to the Common Rule in all matters of research, in all circumstances in which the Common Rule controls. As noted earlier, under the Common Rule, research requiring access to "identifiable" medical records or biological materials falls under the definition of human subject research and is subject to IRB review, which may be expedited. Having said all of that, I must acknowledge that the application of the Common Rule as it now exists to retrospective, non-interactional research requiring access to stored medical information is ambiguous or problematic in several respects.

For one thing, although the AAMC believes that encrypted, linkable medical information should not be deemed "identifiable," the Common Rule clearly does so. We base our position first, on our concern that increasingly strident ideological clashes over privacy may severely obstruct access to "identifiable" medical information, however a federal law or regulation may define "identifiability," and thereby imperil research of great public benefit; and second, on our desire to encourage researchers to use encrypted medical information and spare themselves from the very substantial and costly burdens that new legislation and regulations will inevitably impose upon those requiring access to information that is "identifiable." For the present, access to encrypted, linkable medical information does require IRB review, and we can only hope that IRBs will continue to recognize, as they have historically, that such research poses no more than minimal risk and handle it accordingly.

Second, and much more controversial, is the question of whether and how informed consent should be applied to retrospective, non-interactional research involving the use of archival patient materials. The informed consent requirement has been designed to deal with research that is prospective and involves interaction with human subjects. Indeed, it is difficult to conceptualize what informed consent means for research on stored patient materials that may not take place for months, years or decades. The AAMC consistently has opposed any legislative attempt to mandate a requirement for informed consent for such retrospective research. However, we agree that such research that requires access to "identifiable" patient materials should have IRB review, and we have suggested that the IRB be required to make the following two additional determinations, beyond those already specified in the Common Rule, when evaluating such requests:

  a. Does the research require that the archived patient materials be identifiable? and

  b. Will the research be conducted in a secure environment that will ensure protection of the privacy of the medical information? It is certain that the definition of "secure environment" will be specified in new federal legislation and/or regulation.

Third, under the Common Rule, one of the two most critical determinations that must be made by an IRB is whether the proposed research will expose the research subject to more than minimal risk. The AAMC is very concerned whether this single binary threshold will prove sufficient in dealing with the kinds of complex judgments that will be required of IRBs in considering individuals' "privacy rights" and their possible intrusion. I note that this concern is unique neither to the AAMC nor to this issue; it has earlier been raised before NBAC and State commissions by the mental health and behavioral research communities with respect to the regulation of research involving subjects who are "cognitively impaired," but to no avail. I shall turn in a moment to the Notice of Proposed Rule Making (NPRM) issued in November 1999 by the DHHS[9] to deal with the regulation of medical-information privacy. In that NPRM, the Department has proposed that four additional criteria be added to those already in place for IRB review of research proposals requiring access to identifiable medical information. While objecting to the Department's general approach to this matter because it purposely, and in our view, gratuitously, evades deference to the Common Rule, the AAMC takes particularly strong objection to one of the proposed new criteria that would require IRBs to determine that "(t)he research is of sufficient importance so as to outweigh the intrusion of the privacy of the individual whose information (is to be used)."[10]

We have consistently and strenuously opposed such "weighing" criteria on the grounds that they would force IRB members to render judgments for which there can be no clear-cut or normative standards, and would hold the decisions of the more than 3500 IRBs estimated to exist in the United States today hostage to the personal belief structures, biases, and ideologies of the thousands of individuals who serve on these bodies. The fear of ideological deadlock, akin to what might be anticipated on IRBs containing zealous right-to-life or animal rights advocates, which must review human stem cell or animal research proposals, is not fanciful. There have already been anecdotal reports of its happening in IRBs around the country in their reviews of retrospective studies requiring access to long-stored human tissue samples for purposes of what is deemed by some members of the IRBs to be "genetic research."

Fourth, we must acknowledge the concerns in some quarters that the Common Rule is not sufficiently stringent, or focused, in protecting the privacy of human research subjects and medical information. I would concede that in high-volume research institutions, busy IRBs dealing with the real risks that may attend translational clinical research and early phase clinical trials, may not direct sufficient attention to the assurance of privacy protection. On the other hand, I am not aware that violations of individual privacy have ever been shown to be a significant issue in medical research conducted under contemporary institutional and professional standards. Notwithstanding, at the present time both the NBAC and the Institute of Medicine are engaged in comprehensive reviews of privacy protections of human research subjects and will likely issue reports of their findings and recommendations within the next 12–18 months. In addition, given the imminent relocation and restructuring (and renaming) of the Office for Protection from Research Risks (OPRR) to the Office of the Secretary, DHHS, and the reshaping of its authorities, it is highly likely, in my judgment, that the new Office for Human Research Protection (OHRP) will also undertake such a review as an early order of business and may well recommend changes in the Common Rule.[11]

7. While fully recognizing the political sensitivity of the matter, the AAMC supports the strong preemption of State laws by federal legislation on medical information privacy. The only sytematic exceptions that we would favor would be for those State laws that deal with public health reporting, which tend to be well-integrated into the national reporting network organized and managed by the Centers for Disease Control and Prevention (CDC). We take this

position in recognition of the need for free flows of medical information across state lines in the operations of the health care delivery and payment system, and similarly, in many medical research studies requiring access to large numbers of archived medical records to construct population samples of sufficient size and composition to minimize bias. At the present time, all 50 states have sundry laws that deal with medical information privacy, but there is little uniformity among them, and they are of highly variable reach and effectiveness. In an age of information technology and knowledge-based economies, strongly preemptive federal law would greatly facilitate patient care, the efficient and effective operations of the health-care system, and the conduct of medical research.

8. Finally, the AAMC believes strongly that human subject research databases should be separated by firewalls from clinical databases used in providing patient care, because the research databases can be much more securely protected from trespass than can clinical databases. The argument is based on the plain fact that uninterrupted flows of identifiable medical information are the *sine qua non* for the operations of the contemporary health-care delivery system. Ergo, there is simply no feasible way to restrict access to such medical information to an extent that would come anywhere close to the yearnings of the average citizen; and in fact, none of the recent Congressional bills on this issue, nor the previously noted DHHS proposed rule, would substantially change that fact.

In contrast, it is entirely feasible to approach the publicly desired degree of protection of medical information from unauthorized access with research databases. This assertion rests on the statutory mechanism known as the Certificate of Confidentiality,[12] which, in its 28-year history, has never, to my knowledge, been breached. The Certificate originated in the Comprehensive Drug Abuse

Prevention and Control Act of 1970 (P.L. 91-513) to facilitate the conduct of research into drug trafficking and abuse at the time of the Vietnam War, and in 1988 its protections were incorporated into the Public Health Service Act (P.L. 100-607) and greatly expanded to encompass biomedical and behavioral research more generally. Issuance of the Certificate is by petition from individual investigators for specific research projects, and for most its history has been at the discretion of the Secretary, DHHS, although more recently this authority has devolved to the individual Institutes of the NIH.

The Certificate is issued to protect "sensitive" research information, including "information that if released could reasonably be damaging to an individual's financial standing, employability, or reputation within the community," or information normally recorded in a medical record, the disclosure of which "could reasonably lead to social stigmatization or discrimination."[12] Over the years, the categories of information deemed to be "sensitive" and eligible for protection under the Certificate have significantly expanded. In our contemporary society, the language of the Certificate could not have been better crafted to capture the public's fears about inappropriate disclosure and misuse of any medical information. Put differently, in today's climate of increasing anxiety about the loss of personal privacy, *all* medical information has become "sensitive."

Under the protections of the Certificate of Confidentiality, a researcher can not be compelled to disclose by any federal, state, or local civil, criminal, administrative, legislative, or other proceedings any information from a research database that could be used to identify research subjects. The only statutory exceptions from these sweeping protections are for audits by the funding agency or reporting that is legally mandated to the FDA. There is no other "good cause" exception, and in the only relevant legal challenge to the Certificate's protections, they were determined to be absolute and to cover the entire contents of the research record. It is very important to understand that the Certificate of Confidentiality protects *only* research records and does not extend to information that is considered part of normal patient care and available in the medical record.

For these reasons, we argue that there should be a firewall around research data that clearly demarcates them from patient care records. Only by so doing can patients be assured with certainty that the use of their records and biological samples for medical research will afford them maximum protection of their privacy and expose them to no more than minimal risk—in fact, expose them to far less risk than accompanies the routine use of their records and samples in the provision of medical care. Such a firewall can admittedly pose difficult issues in certain circumstances, perhaps prototypically in the conduct of clinical trials, but I believe these occasions can be managed by carefully crafting research protocols and informed-consent processes. If a clinical trial generates data that will be used in the care of the research subjects, or that have been agreed to be shared with the participants, those data should be entered into the subjects' medical records and handled as clinical information, where the data will be subject to whatever protections and accessibilities pertain to clinical records. In these circumstances, the research database should remain formally and physically distinct from the clinical record.

I will conclude this section by noting that the National Committee on Vital and Health Statistics, which was created in 1996 by the Health Insurance, Portability and Accountability Act

(HIPAA) to advise the Secretary of HHS in analyzing the complex issues of medical information privacy, noted in its report to the Secretary in the summer of 1997[13] that it had been unable to find documentable instances of breaches of confidentiality resulting from researchers' use of medical records.

The Current State of Play

In the last session of the Congress, at least six different bills dealing with medical information privacy were proposed in the House and Senate, from both Republicans and Democrats, and despite intense efforts to build consensus, no bill succeeded in being reported out of committee in either Chamber. The AAMC participated actively in these legislative efforts and worked particularly closely with the staff of Senator Bennett to craft a bill (S. 881; the corresponding House bill was introduced by Rep. Greenwood [H.R. 2470]) that the Association was pleased to endorse. The most intense effort to craft a bipartisan compromise bill took place in the Senate Health, Education, Labor and Pensions (HELP) Committee under the leadership of the chairman, Senator Jeffords, but that effort failed, largely because of two seemingly peripheral issues. One was the issue of parental access to the medical records of minors, which got caught up in the turbulence of abortion politics, and the other dealt with the right of private action, that is, with the right of an individual to seek legal redress for perceived breaches of his/her medical information privacy. The latter issue has proved to be extremely contentious across party lines and has confounded not only medical information privacy legislation, but efforts to pass so-called "Patient's Bill of Rights" legislation, as well. The Senate saga teaches an important lesson about the legislative process: not only are the issues inherent in medical information privacy contentious and challenging enough, but one can never predict when seemingly peripheral issues will suddenly arise to derail the most conscientious efforts to forge political consensus.

With the failure of the Congress to enact federal legislation by August of 1999, the HIPAA mandated that the DHHS enter the fray and issue medical information privacy regulations. Thus the Notice of Proposed Rule Making (NPRM) was published[9] on November 3, 1999, and the comment period, after one extension, closed on February 17, 2000. The NPRM proved to be highly controversial among all stakeholders, satisfying neither privacy advocates, the health care industry, nor academic medicine; the Department reports that it has received 50,000 comment letters, many of them detailed and ranging up to more than 100 pages in length.[14] How the Department will choose to handle this massive volume of commentary is difficult to predict; neither is it possible to anticipate in this presidential election year when the Department will attempt to publish a final rule, nor whether the Congress may be tempted to intervene in response to the deep and vocal unhappiness that the NPRM has elicited across all health-care constituencies.

Space does not permit me to analyze the lengthy, complex, and often highly convoluted reasoning in the NPRM, or to identify the many issues of contention that arise from its specific mandates. I will make the general observation that much of the controversy has arisen from the Department's well-intentioned effort to try to satisfy all of the stakeholders in its articulation of appealing moral principles and promulgation of new patients' rights, and its attempt to implement those principles and rights in rather sweeping, one-size-fits-all fashion across all uses and disclosures of identifiable medical information. In trying to satisfy everyone, it would appear that the Department succeeded in satisfying no one.

Some of the shoals on which this effort seems to have foundered are specific to the NPRM and arise from the Department's persistent efforts to circumvent the limited authority given to it by HIPAA. Thus, HIPAA orders DHHS to promulgate regulations containing standards with respect to the privacy of individually identifiable health information transmitted electronically by health care providers, health plans, and health-care clearinghouses for purposes of standard administrative and financial transactions. Frustrated by their awareness that the problems inherent in medical information privacy substantially exceed this circumscribed ambit, the Department has chosen to extend the reach of the proposed rule to all identifiable medical information that is, ever has been, or ever will be electronically transmitted, thereby embracing essentially all such information, and to embrace not only the three specified classes of covered entities, but all of their "business partners" with whom the entities share such information in the course of conducting their day-to-day activities.

Others "shoals" are not materially different from those that have defied repeated efforts at Congressional resolution over the past several years. These include whether and the extent to which informed consent should be required for accessing, using, and disclosing medical information for various purposes; the applicability of what are often called "fair information practices" to the collection, maintenance, use, and disclosure of medical information; the precise category of medical information that is to be covered by the rule, which centers on very important, and equally contentious, definitions of "identifiability" and "non-identifiability;" the definition of "research" and the requirements specified for accessing medical information for research purposes, both in settings covered by the Common Rule and those that are not; the creation of a federal private right of action by which individuals could seek legal redress for alleged misuses of their medical information; the very controverted matter of how high a threshold (if any) should be established for access to individually identifiable medical information by law-enforcement personnel and agencies; and the impact of the enormous administrative, financial, and operational burdens that would be imposed upon the health care delivery system by the provisions of the proposed regulation. Many of these problems can very aptly be summarized as unintended adverse consequences of lofty goals and well-intentioned aims.

Conclusion

The flows and uses of identifiable patient information in treating patients; paying for their care; operating our extraordinarily complicated system of health-care delivery, financing, and oversight; and conducting medical and health research are bewildering in their complexity and intensity, and exceedingly difficult to explain to the public. At the same time, the increasing public angst about the seemingly inexorable erosion of privacy in our society cannot be ignored, and the ripest target for that angst at this time seems to be medical-information privacy. Polling data reported recently by the California HealthCare Foundation have indicated that as many as 15–20% of the adult populace admit to having paid cash for their medical care to prevent their medical-information from being shared with health insurers, or withholding "sensitive" information from their care-givers, or lying about such information, rather than risk revealing it.[15]

Because the completeness, accuracy, and integrity of the medical record determine its value and define its utility both in caring for patients and performing medical research, the trends suggested by the polling data are troubling indeed and should set off alarm bells in the entire health profession. Yet, the correct "fix" for this problem, one that would restore public confidence without at the same time crippling health-care delivery or hindering, even preventing, essential health research from being performed, continues to elude us. With respect to medical research, the promotion of the health of the public, and access to identifiable medical records, I see the issue as a titanic clash between the assertion of private, sometimes seemingly Hobbesian, rights and concern for the social weal. Such a clash is presently being played out starkly in the State of Minnesota[16] following the enactment there in 1996 and 1997 of seductively simple, but, as it so often happens, dangerously overreaching medical-privacy legislation. It is often stated that in our federal system of governance, the states serve an important function as laboratories in which different approaches to complex public issues can be explored. The report that an AAMC colleague and I prepared following our site visit to Minnesota in 1999 can be found on the AAMC website.[17]

It may well take the proverbial "wisdom of Solomon" to find an acceptable—and workable—resolution of this conflict, but unfortunately, no Solomonic figure has yet arisen. What is, in my view, unarguable is that satisfactory resolution will only be accomplished by carefully crafted federal legislation that creates a comprehensive, nationally uniform, and effective system of workable protections of the confidentiality of medical information. Short of that, recourse to federal regulatory and individual State legislative initiatives will condemn the nation to continued struggling with a patchwork array of discordant and incomplete privacy protections that satisfy no one.

References

[4]Lengauer, C., Kinzler, K. W., and Bogelstein, B. (1997) Genetic instability in colorectal cancers. *Nature* **386,** 623–627.

[2]Alizadeh, A. A., Eisen, M. B., Davis, R. E., et. al. (2000) Distinct types of diffuse large B-cell lymphoma identified by gene expression profiling. *Nature* **403,** 503–511.

[3]Berns, A. (2000) Gene expression in diagnosis. *Nature* **403,** 491,492.

[4]Annas, G. J., Glantz, L. H., and Roche, P. A. (1995) The genetic privacy act and commentary. Boston, MA. Health Law Department, Boston University School of Public Health, Boston, MA.

[5]Murray, T. H. (1997) Genetic exceptionalism and future diaries: is genetic information different from other medical information? in

Genetic Secrets: Protecting Privacy and Confidentiality in the Genetic Era (Rothstein, M. A., ed.), Yale University Press, New Haven, CT, pp. 60–73.

[6]Report and Recommendations of the National Bioethics Advisory Commission (1999) Research Involving Human Biological Materials: Ethical Issues and Policy Guidance, Vol. I., Rockville, MD, August.

[7]Report of the National Bioethics Advisory Commission (2000) Research Involving Human Biological Materials: Ethical Issues and Policy Guidance, Vol. II., Rockville, MD, January.

[8]I am indebted to Mr. Frederick R. Anderson, Attorney at Law, Cadwalader, Wickersham & Taft, Washington, DC, who was, to my knowledge, the first to articulate this insightful distinction.

[9]45 C. F. R. § 160–164. Standards for Privacy of Individually Identifiable Health Information; Proposed Rule. *Federal Register* vol. 64, No. 212, pp. 59918–60065 (November 3, 1999).

[10]Ibid. Part 164.510(j)(3)(vi).

[11]The new Office for Human Research Protection (OHRP) has been established in the Office of the Secretary, DHHS.

[12]Certificate of Confidentiality. 42 U.S.C. sec. 241(d).

[13]Health Privacy and Confidentiality Recommendations of the National Committee on Vital and Health Statistics. Approved on June 25, 1997. Available at http:/www.mssm.edu/medicine/medical-informatics/ncvhs.html.

[14]The AAMC's comment letter on the DHHS Notice of Proposed Rule Making, one of the shorter at only 30 pages, is available at www.aamc.org/advocacy/corres/research/privacy.htm.

[15]Personal communication from Ms. Janlori Goldman, Director, Health Privacy Project, Institute for Health Care Research and Policy, Georgetown University Medical Center, Washington, DC.

[16]Minnesota Statutes 144.335.

[17]McCabe, C. and Korn, D. Confidentiality of Medical Records. AAMC Report on the Minnesota Experience. Available at www.aamc.org/advocacy/issues/research/minrepot.htm.

Abstract

Privacy has served a crucial function in the physician–patient relationship and an essential foundation for maximizing patient health and well-being. The increasing complexity of medical care and financing has weakened the reality of medical privacy as well as patients' perception of confidentiality. In general, computerization has only compounded the problem, making it much easier to multiply and transfer data in ways that stretch the traditional boundaries of confidentiality beyond plausibility.

The primary justification for collecting individual medical data for the purpose of underwriting insurance coverage is insurers' fear of covering a person who knows more about her own risk than the insurer undertaking the coverage. The prospect of genetic tests with the potential to reveal an unprecedented quantity and quality of health information intensifies concern about the conflict between health insurers' approaches to underwriting and patients' and physicians' need for medical privacy. Individual risk assessment for health coverage is inimical to the medical privacy of those who are covered by it. Access to individual patient medical records by health insurers so seriously compromises individual privacy, personal freedom, informed consent, the physician–patient relationship, and access to care that such access may be deemed incompatible with fundamental requirements of an ethically acceptable health-care milieu.

When thinking about the privacy of medical information, we should not conceptualize privacy as an adjunct to health care that is separable and nice, but not essential. Although significant steps have been taken to reduce the incentive of health-benefit providers from using genetic information to restrict coverage, much more needs to be done to ensure the privacy of genetic information.

Privacy and Health Insurance

Can Oil and Water Mix?

Bill Allen and Ray Moseley

The ethical obligation of the physician and other health professionals to maintain the patient's health information *in secret* is both ancient and central to the Hippocratic tradition. Yet the practice has more than a long and distinguished tradition in its favor. Privacy has remained central to this tradition because it serves a crucial function in the physician–patient relationship and an essential foundation for maximizing patient health and well-being. Modern parlance has replaced the Hippocratic language of secrecy with the rubric of confidentiality, which somehow connotatively loses the urgency and intimacy of the notion of secrecy. Perhaps something is also gained, however, since the root concept of confidence points toward an essential correlate, if not a consequence of confidentiality: that of trust.

The increasing complexity of medicine—both its specialization and its social and institutional settings—has significantly threatened and weakened the reality of medical privacy as well as patients' perception of confidentiality. Patients have numerous persons, known and unknown, seen and unseen, who may claim a "need to know" for at least part of a patient's health status and medical history. Under the current division of labor in health care, it may be crucial to good medical care and patient safety for a person in charge of one aspect of care to know about other aspects of care that may be incompatible. Additionally, the barriers have disintegrated between those with a direct patient care interest in having access to at least part of the patient's medical record and those who are concerned with record keeping, scheduling, and the financing of medical care. The reality is that, although many of these "intruders" may have a legitimate need to know only some part of the patient's record, in practice they usually have access to the entire record. Of course, the more persons who have this type of "legitimate" access, the greater the likelihood of a breakdown, whether intentional or not, in the obligation to prevent unauthorized disclosure of sensitive medical information.

The undermining of medical confidentiality was underway before the computerization of medical records and billing became so prevalent. In general, computerization has only compounded the problem by making it much easier to store, to retrieve, to multiply copies, and instantaneously to transfer patient secrets to persons and entities that stretch the traditional boundaries of secrecy, privacy, and confidence beyond plausibility. And we are just beginning to add another layer of multiplicity and complexity by digitalized consults through tele-medicine. The potential for computerized databases and search tools to be used to aggregate otherwise fragmented data also adds threats to privacy that were impossible when such data was too fragmented to enable identification and composite profiling. Even if the security procedures implemented to prevent hacking work, the multiplication of transfers, the possibility of aggregation with other data (both medical and otherwise), and the growth of the volume of medical information collected and stored on patients increase the risk and the magnitude of inappropriate disclosure.

Added to this frontal assault on privacy from the multiplicity of health care providers involved in the actual examination, testing, diagnostic consultations, and procedures are the legion of persons entailed in the financial transactions of billing, auditing, and ultimately paying or reimbursing for the care. At its most extreme, this final insult to medical privacy extends not only to the entity actually named as the primary insurer, but also to the nationally centralized databases of the Medical Information Bureau (MIB) and, potentially, to the international web of stop-loss risk sharing arrangements known as secondary or re-insurance markets.[1] The MIB is an insurance-industry clearinghouse that collects information on insurance applicants submitted by member insurance companies and releases that data to other insurance companies who may be considering the applicant's request for new or increased coverage. The MIB contends that its files do not contain raw medical data, but merely codes noting that some member insurer has declined or restricted coverage based on categories of medical data it ascertained. MIB policy states that other member insurers are not allowed to make underwriting decisions based on the information from the MIB. Rather, the information merely serves as a red flag alerting the insurer considering the application that the applicant has sought coverage before and the category of the data, which may lead the insurer considering coverage to conduct their own investigation or request for information from the applicant's medical record. It is impossible to verify whether this is how the information is actually used.

Health insurance is based on the idea that a group of people, pooling their resources, are better able to protect themselves from potentially devastating costs of health care for illness, disability, or injury. These costs, which may vary widely for any individual, become predictable in aggregate and thus financially manageable. With experience, an insurance company will be able to predict how many individuals in a particular group will develop heart disease, cancer, and so on, and more importantly how much the resulting medical care will cost. This calculation is then translated into the premium charged for each individual in the group. It thus becomes crucial for the successful insurance company to know the health risks of the insured group. The better they can predict these risks, the more competitive, financially stable, and/or profitable the company.

The primary justification for collecting individual medical data, at least for the purpose of underwriting insurance coverage, is that insurers fear covering a person who knows more about her own risk than the insurer undertaking the coverage. For example, insurers worry that a person who knew he had or would develop a disease requiring expensive health care would be tempted to buy insurance coverage he otherwise would not purchase without this "inside" knowledge. Thus, this "adverse selection" might result in the insurer setting the premium at a level that would not cover actual loss.[2]

On a grand scale, if insurers cannot accurately predict the magnitude of risk they will be obligated to cover in claims, they may become insolvent. Some have projected that adverse selection could result not only in insolvency for individual companies, but also in market failure for entire sectors. In large groups seeking coverage, this concern does not require underwriters to collect individual medical information for risk assessment, because the group's average claims experience allows greater confidence in predicting expensive claims, and there is more "room" to spread the costs of the highest claims. Insurers demand more individual approaches to screening smaller groups and individuals for high-risk claims, however, because there is less room to spread the cost of a high claim. Thus fragmentation of the insurance market and screening by the size of employer are said to "necessitate" the collection of individual health data for the individual and small group consumers.

Health insurance is a highly competitive market. One way an insurance company can develop a competitive edge is to identify and to segment individuals with higher risk of costly claims, to improve the claims experience of "healthier" risk pools or to offer lower rates to get new business. After all, if the group is "healthier," the aggregate cost of their health care will be less. This type of segmentation can lead to "death spirals" in which the higher risk groups are eventually unable to afford or even obtain coverage. Competition begins to focus on how to preclude or restrict coverage for the highest risks rather than how to spread the risk over a large group.

This approach to screening individual risk is not some kind of inviolable law of nature, however, and it was not part of health insurance in its early forms. The original health insurance providers charged everyone the same rate, called "community rating," and they covered all applicants who sought coverage. The attempt to carve out healthier groups in order to lower rates for certain employers eventually led to the sort of competition in the late 20th century that resulted in trying to gain business advantage and maximize profits by coverage limitations and exclusions that focused on covering the healthiest consumers and excluding the sickest or at least excluding coverage for their most costly conditions.[3,4] Even the not-for-profit groups, such as Blue Cross-Blue Shield, that began community rating were eventually forced to switch to experience rating and individual risk assessment for individuals and small groups to avert a ruinous competitive disadvantage.

Even as this trend was saturating the insurance market in the late 1980s and 1990s, the prospect of innovative ways of introducing new forms of probabilistic risk information from genetic tests raised significant additional fears. The prospect of genetic tests, developed out of the federal Human Genome Initiative, which had the potential to reveal an unprecedented quantity and quality of health information, intensified concern about the conflict between health insurers' approaches to underwriting and patients' and physicians' need for medical privacy. The advent of predictive genetic tests to indicate risk of common, multifactorial, late onset conditions could easily exacerbate the competition to restrict coverage outlined earlier. Instead of restricting coverage of persons who have already contracted a condition prior to seeking a change of employer or benefit plan, as had previously occurred, predisposition testing could push the competition to restrict coverage or raise rates for persons who have a predisposition, but do not have the medical condition and may never succumb to it. This dynamic would not only increase the barriers to adequate access and coverage, but also the phenomenon of job lock, restricting the ability of persons to seek new employment or advancement for fear of the loss of coverage.

The development of molecular-based genetic predisposition tests for particular conditions are a sufficient reason for concern, but concomitant technological innovations add to the concern. "DNA chips" have provided the means to take a single blood sample and do tests simultaneously seeking markers for numerous genetic susceptibilities. Although such measures may promote efficiency for medical purposes, they also may hinder selective protections of privacy as well as selective disclosure. Most physicians queried by insurance underwriters on specific information, instead of culling the record for specific responses, simply photocopy the entire record and send it to the insurer. Multiplex testing by DNA chips may reinforce the practical difficulties with limiting the scope of disclosure. In addition to genetic markers for common diseases, genetic markers that can distinguish which patients will be allergic or intolerant to particular drugs are rapidly being developed, as well as which regimens may be effective for particular patients with less trial and error.[5] Although this will be a boon to therapeutic interventions, such genetic information could conceivably be used to restrict coverage for those whose conditions may be expected to be uncontrolled.[6]

In light of all this, it is probably not going overboard to conclude that individual risk assessment for health-insurance underwriting is inimical to the medical privacy of those who are covered by it and that the loss of that privacy in the case of those whose coverage is denied or restricted is especially problematic. An important question to be answered is whether the information-gathering aspects of this system are necessary to the functioning of the coverage, and to the extent they are necessary, the scope of what is necessary in the way of compromising privacy. Perhaps even more important is the recognition that the loss of medical privacy compromises good health care. Patients who do not feel they will be able to control the information disclosed in or generated by their encounters with the medical system will withhold information vital to their health care. Moreover, they may be placed in the untenable position of declining to undergo a genetic test that may have a substantial bearing on their lives for fear that they will lose insurance coverage.

Many proposals have been offered to address this privacy issue. The traditional approach is simply to include this information in the consent to treat process. Thus, part of the relevant information a patient would receive before agreeing to a medical treatment is the threat to privacy that may be associated with the proposed medical testing and treatment. This, of course, would provide a patient with information, a necessary ingredient of informed consent. This type of warning is necessary at a minimum, but it does not offer an adequate remedy to the underlying problem. In most cases, merely informing a patient that they may avoid privacy concerns by paying the cost themselves, or by refusing to grant an insurer's request for disclosure of results falsely presumes that a patient has the viable option of not using insurance to pay for their health care. Even the way health insurance is marketed presumes that health insurance is essential and not merely an optional product that one may or may not purchase. Thus, most patients would in essence be financially coerced into "giving" consent.

Since 1991, when we began studying this issue some legislative progress has been made to address this issue. At that time fewer than 10 states had any sort of legal protection against using genetic tests to exclude people from health insurance or to limit their coverage based on genetic status, and those protections were meager. At the present time, a substantial number of states have passed some form of legislation to regulate the use of genetic status for underwriting health insurance. On the federal level, when the Human Genome Initiative began, there was no clear protection against the use of genetic information to restrict coverage for persons seeking health coverage. Although some hoped that the Americans with Disabilities Act (ADA) might provide some protection against genetic discrimination, there was considerable controversy and uncertainty about whether genetic predispositions to disease would qualify as protected disabilities. Even if genetic predispositions did qualify for ADA protection, clear language in the final chapter of the act excluded underwriting of health benefits from discrimination otherwise prohibited by the Act. In recent years that lack of protection has been partially remedied by inclusion in the Health Insurance Portability Act (HIPA), which includes language prohibiting exclusion of health care benefits based on genetic information for persons who are covered by the act.

The state regulations for health-benefit protection against genetic discrimination do not provide an adequate remedy either. For one thing, state insurance laws only protect persons whose health benefits are provided by conventional insurance compa-

Although this progress is commendable and offers significant protection to many, it has not solved the problem. To begin with, the federal HIPA does not cover everyone. It only covers persons who were already protected under their employer's health benefit plan when they change to another employer that provides health coverage. It does not protect people who do not already have coverage. Nor does it protect those who leave an employer for a new employer where no such coverage exists. As recent news stories have continually documented, the number of the uninsured has increased in the last decade of the 20th century, even as the economy achieved a record sustained expansion.[7]

The state regulations for health-benefit protection against genetic discrimination do not provide an adequate remedy either. For one thing, state insurance laws only protect persons whose health benefits are provided by conventional insurance companies, which are subject to state regulation. They do not protect the huge numbers of persons whose health-benefit coverage is underwritten by a self-insured employer, union, or other entity, which are not generally exempt from state regulation.[8] Moreover, such protection as is afforded by these state laws is uneven and inadequate. A few more than half the states prohibit the use of genetic tests to refuse issuance or renewal of coverage. A few less than half, however, also prohibit health insurers from charging higher rates or premiums on the basis of genetic tests or conditioning terms or disbursement of benefits on genetic test results.[8]

Access to individual patient medical records by health insurers so seriously compromises individual privacy, personal freedom, informed consent, the physician–patient relationship, and access to care that such access may be deemed incompatible with fundamental requirements of an ethically acceptable health-care milieu. Some have suggested that separate medical records for genetic test results would be a way to deal with this problem. This solution is unworkable, however, because it fragments the record and creates practical problems, including the risk of omission of factors crucial to diagnosis, intervention, or safety (such as drug intolerance), or confusion between patients' genetic profiles. A medical record is an integrated whole, and omissions would create gaps that can only be explained by inferences concerning genetic tests and interventions based on them. Insurers' expert reviewers would be likely to recognize such redacted records and guess at the meaning. Whether their guesses were correct or not, the damage to patients would be done.

Other suggestions have included anonymous genetic testing. This is also impractical. This would magnify insurers' concerns about the risks of adverse selection, but would also not avoid problems detailed earlier. The value of knowing one's genetic status would be severely limited if she were unable to disclose it to her physician without that being tantamount to disclosure to the insurer. The answers seem rather to lie in a different direction.

Part of the loss of medical privacy for patients has been the involvement of third-party payors. Payors argue that the stake held by employers, insurers, and other insured parties entitles them to restrict the informational privacy of the medical encounter. Privacy for medical information may not be an absolute value, trumping all other claims of a right to access patient information. Yet privacy concerning such intimate details as health status and the personal information that is inevitably enmeshed in it remains such a primary value that it should trump all but the most compelling interests of third parties. When individual medical privacy is curtailed on the basis of such a compelling interest, the scope of the exception should be narrowly tailored and vigorously enforced. Among the justifiable exceptions might be the prevention or detection of fraud. Using genetic information to exclude or restrict coverage for at risk persons, however, is not such a justifiable exception. To the extent that the potential for less predictable large claims with less ability to spread such losses is a feature of small employers and individuals, risk pooling cooperatives should be instituted to mitigate this problem, so that individual risk screening is not needed. The historical accident that employer-based insurance yields a variety of sizes of the groups seeking coverage is not an adequate justification for a lower standard of medical information privacy for small groups or individuals.

When thinking about the privacy of medical information, we should not conceptualize privacy as an adjunct to health care that is separable and nice, but not essential. Privacy of medical information should be understood as part of the service provided. That means third parties, such as government or commercial health-benefit purveyors, are not really primary stakeholders in individual medical information. They may have legitimate interests in aggregated anonymous information about the delivery and the aggregate risk of the group, but they should provide it without access to individualized identifiable data.

There are a number of reasons why prohibiting access to genetic information by insurers is unlikely to result in adverse selection sufficiently substantial to cause insolvency or market failure. To begin with, people seek health coverage for other reasons than genetic predispositions to disease. For example, a woman who discovered she did not carry the BRCA1 gene would still be at the same substantial risk of breast cancer from non-heritable causes as all women. Thus, women who test negative for BRCA1 are not likely to drop health insurance coverage, thereby leaving only "high risks" in the insurance pool. Moreover, because genetic predispositions to most common diseases do not allow for prediction of severity or age of onset, consumers could not successfully game the system by waiting until just before onset of symptoms to acquire coverage. Unlike life insurance, health benefits do not come in face value amounts, so that persons who have a positive test can dramatically increase their coverage.

For self-insured employers, as far as health benefits alone are concerned, the same arguments apply. But much more is at stake in the case of employer access to medical information about individual employees. In addition to using genetic information to cut health-care costs, employers may use the information as a basis for promotion, dismissal, and other crucial decisions. Protections for privacy must especially be strengthened for employees.

None of the aforementioned rationales for insurer or employer access to individual medical information is compelling enough to overcome the rights of patients to the privacy of their medical information. The direction of the protections encompassed in state laws and HIPA restricting the use of genetic information for health-coverage underwriting should be extended to cover everyone in order to remove any incentive to use genetic information to restrict coverage. Finally, the type of protection afforded by the ADA for protected disabilities should be strengthened and enforced to prevent employers from genetic discrimination in other areas besides health coverage.

References

[1]Moseley, R., Allen, B., McCrary, V., Dewar, M., Crandall, L., and Ostrer, H., (1995) The Ethical, Legal and Social Implications of Predictive Genetic Testing for Health, Life and Disability Insurance: Policy Analysis and Recommendations, University of Florida.

[2]Pokorski, R. (1992) Use of genetic information by private insurers: genetic advances: the perspective of an insurance medical director. *J. Insur. Med,* 2460–2468.

[3]Kolata, G. (1992) New insurance practice: dividing sick from well. *The New York Times*, March 4, p. A1, A7.

[4]Quinn, J. B. (1993) Insurance: the death spiral. *Newsweek,* February 22, p. 47.

[5]Kolata, G. (1999) Using gene tests to customize medical treatment. *The New York Times,* December 20, p. A-1.

[6]Arthur, C. (1999) DNA chip testing could be abused. *The Independent* (London), April 9, p. 7.

[7](2000) On providing access to health care for all our citizens. *St. Louis Post-Dispatch*, January 2, p, B-1.

[8]Mariner, W. K. (1992) Problems with employer-provided health insurance: the employee retirement income security act and health care reform. *New Engl. J. Med.* **327,** 1682–1685.

Abstract

Individuals and organizations have adopted computers so quickly and in such large numbers that they have paid insufficient attention to the moral implications of such rapid change. As more and more personal data are kept on computers and disks, increasing numbers of people have authorized, as well as easy unauthorized access. Rules that ought to govern the handling of these data are often left unspecified. Furthermore, different organizations differ in their rules, because no consensus has yet been reached on issues of privacy and security. The need for clear moral thinking and moral argument is therefore especially important in this area.

Every individual and organization has an obligation to respect privacy, but privacy is a complex phenomenon. There are various kinds and types of privacy. The meaning of the term is always dependent on the context of its use. Moreover, there is no consensus among philosophers or lawyers on how best to define privacy, or how far privacy ought to extend, or how to balance a concern for privacy against other moral considerations. Discussions that approach privacy as if there were a distinct and coherent definition are simplistic. If we are interested in rational evaluation, we should keep straight the kinds and types of privacy and the principles applicable to particular issues or events.

In this chapter, I begin a consideration of the nature and limits of medical-information privacy with an analysis of the meaning of words. Analytic investigation is an important tool in the evaluation of privacy. Once I analyze and clarify the terms of the argument, I then proceed to an evaluation of medical-information privacy and how to balance privacy against other moral considerations. I conclude with an evaluation of the application of data mining algorithms to patient information and medical records.

Data Mining, Dataveillance, and Medical Information Privacy

Mark E. Meaney

Introduction

Individuals and organizations have adopted computers so quickly and in such large numbers that they have paid insufficient attention to the moral implications of such rapid change. As more and more personal data are kept on computers and disks, increasing numbers of people have authorized, as well as easy unauthorized access. Rules that ought to govern the handling of these data are often left unspecified. Furthermore, different organizations differ in their rules, because no consensus has yet been reached on issues of privacy and security. The need for clear moral thinking and moral argument is therefore especially important in this area.

Every individual and organization has an obligation to respect privacy, but privacy is a complex phenomenon. There are various kinds and types of privacy. The meaning of the term always depends on the context of its use. Moreover, there is no consensus among philosophers or lawyers on how best to define privacy, or how far privacy ought to extend, or how to balance a concern for privacy against other moral considerations. Discussions that approach privacy as if there were a distinct and coherent definition are simplistic. If we are interested in rational evaluation, we should keep straight the kinds and types of privacy and the principles applicable to particular issues or events.

In this paper, I begin a consideration of the nature and limits of medical-information privacy with an analysis of the meaning of words. Analytic investigation is an important tool in the evaluation of privacy. We can properly examine arguments for or against certain uses or functions of privacy only if we first get some clarity on the relevant concepts. Analysis can uncover the presuppositions of different positions and so evaluate them for consistency, clarity, and applicability. Once I analyze and clarify the terms of the argument, I then proceed to an evaluation of medical-information privacy and how to balance privacy against other moral considerations. I conclude with an evaluation of the application of data mining algorithms to patient information and medical records.

The Concept of Information

It should come as no surprise that philosophers and lawyers have not yet reached consensus on how best to define privacy. While agreement on a distinct and coherent definition is improbable, all agree that the word has several meanings and uses depending on the context. Ordinary usage reflects the wide range of our experience as well as the fuzziness and ambiguity of the term. "Privacy" has at least four different meanings:

1. Privacy of the person;
2. Privacy of personal behavior;
3. Privacy of personal communications; and
4. Privacy of personal data.

The term can have any or all of these meanings depending on the context of its use.

For instance, sometimes we use the term privacy to refer to privacy of the person, or the individual. Here we usually mean that people value autonomy, or the exercise of control over mind and body. If we are to function as autonomous or self-governing agents, we need to keep certain areas of our lives free from interference. For example, in her article, "Privacy and the Limits of Law," Ruth Gavison analyzes the meaning of privacy in terms of solitude and intimacy.[1] Solitude permits personal reflection on ends and means in the course of one's life, while intimacy involves close, relaxed, and sincere relations between two or more persons. Individuals in an intimate state trust each other completely. They are therefore willing to share advice and to help one another make decisions, and so on. For Gavison, solitude and intimacy are essential to the development of the person. In her view, privacy limits access to a person.

A different but related use occurs when we refer to privacy of the person in spatial terms. The earliest application of privacy in US courts was place-oriented.[2] The fourth amendment of the US Constitution guarantees the right of people to the security of their "persons, houses, papers, and effects," against unreasonable searches and seizures. The amendment describes privacy of the person in terms of a private place, a secure environment, or private property.

We also use the term privacy in relation to behavior. Our autonomy depends on our ability to make choices free from the illegitimate influences of others. The privacy of personal behavior refers to all aspects of personal choice, but especially to sensitive matters, such as habits, religious practices, and political activities, both in private and in public places. In his work *Privacy and Freedom,* Alan Westin analyzes privacy in terms of the more fundamental notions of "reserve" and "anonymity."[3] He describes reserve as a psychological barrier built by a person for his or her own personal reasons, and anonymity as freedom from identification, surveillance, or intrusion. In Westin's view, privacy provides a zone of protection so that persons can choose to act for themselves by considering alternatives and consequences. Privacy therefore refers primarily to an interest persons have to determine for themselves the extent to which others have information about their personal activities and/or behaviors.

Common usage of the word privacy refers also to personal communication. Personal communication can take the form of various kinds of emotional release, "confidences" of different sorts, and other kinds of private and public discourse. People desire privacy of personal communications so that they can feel confident that others will respect disclosure of personal matters. We have an interest in being able to communicate freely, using various media, without illegitimate monitoring of our communications by other persons or organizations. In *"Privacy: A Moral Analysis,"* Charles Fried analyzes privacy not simply in terms of limits on information about us, but in terms of the measure of control we have over the quality of the information others have about us.[4] In Fried's analysis of the concept, privacy refers to an interest people have in controlling the quality of the information that is communicated about them to others. On this view, privacy of personal communications serves as the basis of the legally protected right against forced disclosure in certain relationships, e.g., health professionals and their patients, lawyers and their clients, clergy and their parishioners and so on.

A more recent use of privacy refers to the storage of personal data. The relation of computing to communication since the 1980s has served closely to link a concern about the privacy of personal communications to a concern about the privacy of personal data. In Roger Clarke's work on privacy, for example, "information privacy" refers both to the privacy of personal communications and of personal data.[5] In the body of his work, Clarke argues in favor of a strong version of information privacy. For Clarke, information privacy is an interest people have that data about themselves should not be automatically available to other individuals or organizations. Even where third parties possess data, Clarke argues that people ought to be able to exercise a substantial degree of control over all data about themselves. Thus, information privacy is the interest a person has in controlling, or at least significantly influencing, the handling of all data about themselves. I refer to Clarke's claim as strong privacy—the strong version of information privacy. Strong privacy, then, is a claim covering all data about an individual or a group.

Clarke's defense of strong privacy suffers from a lack of clarity in his use of terms. A failure to get clear on the meaning of terms results in a failure on his part to determine correctly how far information privacy ought to extend and how to balance a concern for information privacy against other moral considerations. For example, he does not properly analyze the concept of information. He simply states that the term "information" "implies the use of data by humans to extract meaning," while the term "data" refers to "inert numbers."[6] In short, information is simply data that has value. In what follows, I argue that, because Clarke does not consider other meanings of information, his analysis of information privacy is also incomplete. Strong privacy is flawed at the root.

Richard De George describes in his book *Business Ethics* the ordinary use of words in discussions about computers.[7] The meaning of the word information is often left ambiguous. In an analysis of the concept, De George warns that we should be careful to distinguish "facts," "data," "knowledge," and "understanding" in order to be clear on the meaning of the terms.[8] Just as Eskimos have many different words to describe kinds of snow, so we can use a variety of different words to characterize different kinds of information. Moreover, how we use the term has moral import.

By "fact" we mean a statement of the way the world is independent of our knowledge of it. "Knowledge" in part consists in facts as known. "Understanding" consists of knowledge that is integrated in some unified way and evaluated. Facts, knowledge, and understanding are public in that the individual appropriation of facts does not deprive anyone else of them. Information in this sense is "infinitely shareable."[9] We can all know the date of the discovery of America, or that five times five equals twenty-five. Such knowledge on the part of one person does not prevent others from knowing these facts.

When we speak of "information systems," on the other hand, we often use the word information to cover data, facts, and knowledge. Strictly speaking, however, while we may store interpreted facts informed by theories in a hard drive, systems do not know the difference. They simply contain symbols, or as Clarke says, "inert numbers," that can be interpreted by those who know. We refer to these symbols as "data." What we enter into a computer is data. The entered data can represent words, letters, or numbers, and so the data can represent facts in coded fashion. But data can represent falsehood as well as facts, so it is important not to conflate data, facts, and knowledge in our use of the term information.

Much hinges on De George's analysis of the concept of information. His distinction between data and facts separates the private from the public, or what individuals and organizations can own and what they cannot. De George argues that individuals and organizations cannot own facts that are available to all, but they can own data representing facts.[10] Facts entered into a computer as data belong to whoever owns the computer. Individuals or organizations own data in the sense that they have the right to exclusive use of the data entered into their computers. If they use the data to produce something else, that product in turn also belongs to them. As property, they have the exclusive right to manage the data, the right to the income from the use of the data, the right to replace or erase it, the right to keep it indefinitely, and the right to transfer it to other computers.

Of course, part of the notion of private property also involves liability for its use. De George notes that one cannot simply use one's property in any way one wishes. If, for example, one owns a gun, one cannot legitimately use it to harm or threaten harm to others. The same is true of data. Just because someone owns data does not give him the right to use the data any way he wants. Thus, so long as the owners of data harm no one and threaten no harm, then data are the owners to use as they wish.

De George's analysis of the concept of information is important to an evaluation of information privacy. Keeping facts and data distinct helps us to state, and so resolve, a number of important issues. How one analyzes the relevant concepts determines in large measure one's claims about how far information privacy ought to extend and how to balance a concern for privacy against other moral considerations. For example, Roger Clarke's analysis of information and information privacy is incomplete, because he fails to keep distinct facts from data. Clarke's description of information, in turn, determines in large measure his analysis of information privacy. In short, his analysis of the meaning of the words has direct bearing on his claims about how far information privacy ought to extend and how to balance a concern for privacy against other moral considerations. In the next section, I argue that Clarke's strong version of information privacy is too broad in scope and application, primarily because he fails to keep distinct facts and data. De George's analysis of information suggests a more moderate approach to information privacy.

The Nature and Limits of Information Privacy

For De George, facts are simply statements about the way the world is independent of our knowledge of it. By definition, information in this sense is public and not private. No one can lay claim to facts that are available to all. Furthermore, if we cannot own facts, then we cannot own facts about ourselves, any more than we can own other kinds of facts. Facts about us do not belong to us. This does not mean, however, that everyone has a right to know all the facts about us. It simply means that, if we cannot own facts, information privacy cannot extend to all facts about us.

For example, suppose that the statement "Jill subscribes to a certain magazine" is a fact. Jill cannot lay claim to this fact; she does not own this fact. She knows she subscribes to the magazine, the friends she tells about her subscription know, employees of the magazine know, the mail carrier knows, others who might see the magazine in her mailbox know, or the administrative assistant who distributes the mail in her office knows, and so on. Suppose further that the publisher of the magazine has entered this fact into its computers. Jill's name now appears on a list of all subscribers to that magazine as data. Does the publisher have the right to reveal to others that Jill is a subscriber; does the publisher have the right sell a list with Jill's name on it?

Although no one can own the fact of Jill's subscription, others can own data representing this fact. The publishing company has the exclusive right to manage data entered into their computers, the right to the income from the use of the data, the right to replace or erase it, the right to keep it indefinitely, and the right to transfer it to other computers. In short, the publisher has the right to sell the list, unless selling the list to another magazine for promotional purposes, or to some other advertiser, would somehow harm Jill or threaten harm.[11] An increase in the amount of mail in her mailbox, however, does not constitute harm. She may suffer an inconvenience, but it does her no harm.

Should information privacy extend to data about Jill of the sort "she subscribes to a magazine"? Strong privacy entails it does. Roger Clarke argues that information privacy is a claim that individuals have on other individuals or organizations that data about themselves should not be automatically available to other individuals or organizations. Moreover, even where other individuals or organizations possess such data, Clarke argues that persons must be able to exercise a substantial degree of control over that data and their use. De George's distinction between facts and data shows that Clarke cannot be correct about the extent or scope of information privacy. The strong version of information privacy cannot be right. Jill does not own facts about herself of the sort "she subscribes to a magazine," whereas the publisher can own this fact as data. So long as owners of data cause her no harm or threaten no harm, then data they own are theirs to use as they wish. They do not violate Jill's information privacy by selling a list with her name on it without her knowledge or consent.

Does it make any difference if data from the database containing Jill's subscription are combined with data from a database containing data that Jill pays her bills on time and that she buys certain products? Suppose a firm specializes in data warehousing; they own and operate a data mart. They acquire data from multiple internal and external sources, and manage their data in a central, integrated repository. The firm provides analysis tools to interpret selected data, and produces corporate reports to support managerial and decision-making processes. Suppose further that a marketing agency wants access to the repository and hires a firm to extract valuable marketing information. The firm "leases" the data mart and applies a data-mining algorithm to the prepared data. The algorithm compresses and transforms the data so that the marketing agency can easily identify latent, valuable marketing information. The data mining process yields the following information about Jill: (1) she is a 29-year-old single, Asian-American female; (2) Vice President of operations for a hospital with a salary of approximately $110,000; (3) lives in downtown Atlanta; (4) buys a certain kind of magazine; and (5) has recently purchased a new oven for her new condo with a platinum Visa card. Moreover, the data mining algorithm yields the conclusion that she will likely purchase a blue blazer in the next six months because she's a member of a group with similar characteristics. The marketing firm mails out a coupon for a blue blazer.

Some argue that, although it might be permissible to keep records about us in various places, the ability to gather all of these data in a data mart through data-warehousing technology somehow violates our privacy. This argument is mistaken. Facts about us do not belong to us. If those who have facts about us stored as data have the right to sell the data, then warehousing data does not violate our privacy, so long as they do no harm or threaten harm to us. Of course, ownership does not confer on the marketing agency the right to use data to harm Jill, or anyone else for that matter. If the marketing agency used data in ways that harmed Jill, she would have the right not to have these data given to others.

The Nature and Limits
of Medical-Information Privacy

Clarke fails to analyze properly the concept of information. His failure to get clear on the meaning of the term affects adversely his analysis of information privacy. Under the terms of strong privacy, individuals must give consent to the use or dissemination of all data about themselves. De George's analysis of the concept of information shows that information privacy cannot cover all facts about us. Moreover, Clarke's strong version of information privacy leads him to the following conclusion. If individuals and organizations avail themselves of any data about other individuals without their express knowledge and written consent, then this unauthorized use of personal data constitutes a form of data surveillance, or so-called *"dataveillance."*[12]

The Oxford Dictionary defines surveillance as "watch or guard kept over a person, esp. over a suspected person, a prisoner, or the like; often spying; supervision for the purpose of direction or control."[13] *Webster's 3rd Edition* defines surveillance as "1. Close watch kept over one or more persons: continuous observation of a person or area (as to detect developments, movements or activities); 2. Close and continuous observation for the purpose of direction, supervision or control."[14] "Surveillance," then, is the systematic investigation or monitoring of the actions or communications of one or more persons to collect information about them, their activities, or their associations in order to direct or control their behavior.

Clarke defines dataveillance as the systematic use of data systems to collect information in order to direct or control behavior.[15] Dataveillance differs from surveillance insofar as dataveillance is not applied to real individuals or groups, but to the "data-shadows," the "data trails" or the "digital persona" of real individuals or groups. Dataveillance comprises a wide range of techniques, including the integration of data, screening or authentication of transactions, front-end verification, front-end auditing, cross-system enforcement, computer matching, and computer profiling.

Clarke distinguishes between "personal dataveillance" and "mass dataveillance." Personal dataveillance is the systematic use of personal data systems in the investigation or monitoring of the actions or communications of an identified person as a means of controlling the individual's behavior. Mass dataveillance is used to direct or control group activity or behavior. Dataveillance is "by its very nature, intrusive and threatening."[16] Thus, Clarke's strong version of information privacy entails that use or dissemination of any data about individuals without their knowledge and consent amounts to a kind of surveillance, and therefore constitutes a violation of information privacy.

De George's claims suggest a more moderate approach. His definitions of fact and data are compatible with several restrictions on the use of data and on the method of learning facts. Although no one owns facts about us, it does not follow that all facts about us are properly common knowledge. The statement that no one owns facts does not mean that everyone has a right to know all facts.

Suppose that the statement "Jill had an operation last April" is a fact. Just as Jill cannot own the fact of her subscription, she cannot own this fact either. Jill knew she was in the hospital; the physicians, nurses, and allied health professionals in attendance knew; the billing staff at the hospital knew; employees of her health maintenance organization knew; the friends and family she told about the operation knew, as did anyone else Jill told. However, although no one owns the fact that Jill had an operation, it does not follow that her operation is properly common knowledge like the distance between Boston and New York. Nor does the statement that no one can own facts mean that others have a right to know this fact.

Our society recognizes certain areas of privileged communications.[17] Physician–patient confidentiality, lawyer–client confidentiality, and priest–penitent confidentiality are all examples of privileged communications. Facts revealed in such disclosures are confidential. If society did not respect the privacy of certain personal communications, this would seriously impair the practice of medicine, law, and so on in well-known and documented ways.

Now, suppose further that billing clerks entered the fact of Jill's operation into her HMO's database. The HMO would then own the data representing the fact of Jill's operation. Limits to the use of such data come not only from liability, but also from the strictures of physician–patient confidentiality. Even though Jill's HMO owns the fact about her operation as data, this does not give the HMO the right to use the data any way they want. Her HMO may own the data in the sense that they have the right to exclusive use of the data as put into their computers, but they cannot use the data in ways that would harm or threaten harm. The HMO cannot, for example, violate the confidentiality of the physician–patient relationship. Data are the owners to use as they wish, so long as they harm no one and threaten no harm.

A clear example of a violation of information privacy occurred recently after Giant Foods and CVS Pharmacies entered into an agreement with several pharmaceutical firms. In a joint venture, the partners hired Elensys, Inc. to mine data in the CVS database. The database contained information sufficient to fill the prescriptions of all their customers, both past and present. The mining process generated a letter with the pharmaceutical firm's letterhead, which advertised a product depending on the kind of the prescription. For example, one such letter began,

> "Our records indicate that you have recently sought treatment for sexual dysfunction by using a prescription product.

We hope you have been successful in your treatment, but if
you are still experiencing difficulties, we have good news
for you. Recently, a new product was introduced . . ."

De George's position entails that, although no one owns
facts about the state of affairs of patient diagnosis and treatment,
CVS Pharmacies does own the data about customer prescriptions. Nevertheless, CVS does not have the right to allow pharmaceutical companies unauthorized assess to personal data for
marketing purposes. A gross violation of the privacy of personal
communications, the letter was both intrusive and threatening.

Others have written extensively on how breaches of confidentiality undermine the physician–patient relationship, so I won't
belabor the point here. I do wish to explore, however, the extent
to which information privacy should extend to facts about Jill of
the sort "she had an operation in April," when individuals or
organizations have stored such facts as data.

Suppose Jill's HMO wants to engage in outcomes measurement (OM) to improve the quality and efficiency of the delivery
of care. A form of data mining, outcomes measurement involves
the examination of clinical encounter information, insurance
claims, and billing data to measure the results of past treatments
and processes.[18] Outcomes measurement can serve a variety of
purposes. Suppose that Jill's HMO wants to share its studies to
help providers cut costs and improve care by showing which
treatments statistically have been most effective. The ethical issue
concerns the extent to which the HMO must secure Jill's consent
to use and disseminate data on her operation.

Strong privacy entails that the HMO cannot use and disseminate any data about Jill's operation without her express knowledge and written consent. De George's position suggests a more
moderate approach. I argued earlier that the publisher of Jill's
magazine did not need Jill's consent to sell a list with her name
on it. Moreover, the marketing agency did not need Jill's consent
to reach the conclusion that they should send her a coupon for a

blue blazer. The HMO does own facts about Jill's operation stored as data. They cannot, of course, use the data any way they wish, because they do not have a right to harm Jill. The question remains, does medical information privacy cover all facts about Jill's operation stored as data, or can Jill's HMO use and disseminate some data without her consent?

In his article "Medical Records: Enhancing Privacy, Preserving the Common Good," Amitai Etzioni develops a number of categories that I use to address these questions.[19] Etzioni distinguishes among an inner circle, an intermediate circle and an outer circle of personnel directly or indirectly involved in the delivery of health care. Inner circle personnel are all those directly involved in the treatment of the patient. The intermediate circle includes health-insurance and managed-care corporations. Outer circle personnel include parties not directly involved in health care, such as life insurers, employers, marketers, and the media. I use these categories to make the following claim.

A complete interpretation of meaning of the statement "Jill had an operation in April" involves multiple layers of different kinds of facts. Following Etzioni, I define inner-circle facts as personal communications between the patient and health care professionals, or communications among those directly involved in the treatment of the patient. Medical history, treatments, and lab results are all examples of inner-circle facts. Inner-circle data are those kinds of facts as stored. Intermediate-circle facts include such facts as outcomes, utilization of resources and methods of reimbursement. Likewise, intermediate-circle data are those kinds of facts stored as data. Outer-circle facts and data would then include facts and data not directly related to the delivery of health care.

In outcomes measurement, Jill's HMO would need to secure her consent only for use and dissemination of inner circle facts and data. Drawing on the work of Gavison,[20] Westin,[21] and Fried,[22] Jill's HMO ought to respect the privacy of her person, of her personal behavior, and of her personal communications. Moreover, the confidentiality of patient information protects both inner-circle facts and data. Consequently, Clarke's strong version of information privacy does apply to inner-circle facts and data. Individuals and organizations should not automatically have access to these sorts of data. Also, Jill should be able to exercise a substantial degree of control over all inner-circle data.

However, Jill cannot lay claim to intermediate-circle facts any more than she can lay claim to other kinds of facts. On the other hand, her HMO does own the facts of her operation stored as data. The HMO has a right to the exclusive use of these data, so long as such use and dissemination does not harm Jill or threaten harm. By definition, intermediate-circle data are cleansed of personal communications. The HMO does not, therefore, violate Jill's right to privacy when they make use of such data in outcomes measurement without her knowledge and consent. In short, medical-information privacy ought not to extend to intermediate-circle data.

For example, Jill cannot lay claim to facts of the sort that a certain kind of treatment had a specific outcome, or that her physician on average refers patients to x number of specialists, or that a prescribed drug cost x amount. Consequently, her HMO does not need to inform her of outcomes measurement applied to the analysis of intermediate circle data. If outcomes measurement does not require access to inner-circle data, it does not require informed consent. As regards outer-circle facts and data, if, for example, Jill is a member of a certain class of patients, her HMO could sell the characteristics of membership in that class to a marketing agency. The marketing agency could then use the demographic profile legitimately to market products. Thus, while information privacy ought to protect Jill's personal communications, it ought not extend to what I've called intermediate- and outer-circle data. Individuals and organizations need Jill's permission to use and disseminate inner-circle data, but they do not need her permission to mine intermediate- and outer-circle data.

De George's analysis also allows us to reach the following conclusions. By definition, facts are true, because they are statements about the way the world is. Data are neither true nor false. Entered data can represent words, letters, or numbers, and so the data can represent facts in coded fashion. But data can represent falsehood as well as facts, and when interpreted, data may yield falsehoods. This may be either because clerks entered falsehoods rather than facts into the computer as data, or because they made mistakes in entry, or because the software was defective, or for a variety of other reasons.[23] There is no guarantee that the data represent facts.

It follows that, although Jill's HMO owns the data, it is responsible for the interpretation and use it makes of the data, i.e., harm done by mistakes, computer glitches, and so on. Consequently, even though the HMO may not violate Jill's privacy in outcomes analysis of intermediate- and outer-circle data, the use and dissemination of data may cause Jill harm if she does not have the opportunity to verify what the data contains, to correct errors, or to rebut false statements. Jill's right to inspect her personal data is an instance of a general right against credit agencies, governments, and any others who maintain and use data in ways that directly affect, and could harm her. Thus, although her informed consent may not be required for particular purposes, she should be notified of any such files kept on her, and she should have the opportunity to correct errors or to rebut false statements.

Conclusion

In a consideration of the nature and limits of medical-information privacy, I have argued that we must first get some clarity on the relevant concepts, specifically on the meaning of information. Clarity about the meaning of terms is necessary in order to evaluate the basis and legitimacy of the application of information privacy. Through De George's analysis, we were able to uncover the presuppositions of Roger Clarke's position, and so evaluate the strong version of information privacy for consistency, clarity, and applicability. Clarke failed to analyze information adequately in relation to facts and data. This lacuna undermined the viability of strong privacy, its overall coherence, and its ability to handle and account for several morally relevant facts.

De George's analysis of the concept of information shows that information privacy cannot extend to all facts about us that might be stored as data. The use or dissemination of data about individuals without their knowledge and consent is not necessarily a violation of their information privacy. Thus, data mining is not by its very nature intrusive and threatening. Nevertheless, the right to exclusive use of data does not confer on owners a right to harm or threaten harm to others. Limits to use and dissemination of data stem not only from liability for use, but also from the strictures of confidentiality. Simply because no one can own facts does not mean that everyone has a right to know all facts. We may have no right to facts, but we also have no general obligation to make known personal facts about ourselves. When we make personal facts about ourselves known to other in confidence, we have the right to insist and expect that they will be kept confidential. Society recognizes certain areas of privileged communications, and for good reasons. Owners of data therefore have an obligation to determine that confidential records are kept secure.

On the other hand, those of us living in the so-called information age ought to be aware of a distinction between facts and data, i.e., between what is properly public information, or common knowledge, and what is private. Some facts about us that are stored as data are facts that are available to all. Individuals and organizations can therefore use and disseminate these data without our knowledge and consent. Although we may have a right to correct errors and rebut false statements, our right to privacy may be limited by the property rights of the owners of data. Ethical and legal analysis in the future will undoubtedly focus on the categories that distinguish kinds and types of data. Such analysis ought to sort data according to inner-, intermediate-, and outer-circle categories in order to delimit the private and the public.

References

[1]Gavison, R. (1980) Privacy and the limits of the law. *Yale Law J.* **89,** 421–471.

[2]McCloskey, H. J. (1980) Privacy and the right to privacy. *Philosophy* **55,** 17–38.

[3]Westin, A. (1967) *Privacy and Freedom.* Atheneum, New York, p. 22.

[4]Fried, C. (1980) Privacy. *Yale Law J.* **77,** 475–493.

[5]Clarke, R. (1991) Information Technology and Dataveillance. http://www.anu.edu.au/people/Roger.Clarke/DV/CACM88.html>pp. 1–20.

[6]Clarke, R. (1993) Introduction of Dataveillance and Information Privacy. http://www.anu.edu.au/people/Roger.Clarke/DV/Intro.html>p. 3.

[7]De George, R. T. (1995) *Business Ethics.* Prentice Hall, Englewood Cliffs, NJ, pp. 336–360.

[8]De George, pp. 347–349.

[9]De George, p. 356.

[10]De George, pp. 347–348.

[11]De George, pp. 357.

[12]Clarke, pp. 10–13.

[13]Oxford Dictionary (1993) vol. X p. 248.

[14]Webster's 3rd Edition (1976) p. 2302.

[15]Clarke, pp. 6–8.

[16]Clarke, p. 3.

[17]De George, pp. 356–357.

[18]Bresnahan, J. (1997) Data Mining: A Delicate Operation. *CIO Magazine,* pp. 1–8. <http://www.cio.com/archive/061597_mining_content.html>

[19]Etzioni, A. (1997) Medical records: enhancing privacy, preserving the common good. *Hastings Center Report* **29(2)**, 14–23.

[20]Gavison, pp. 421–471.

[21]Westin, p. 22.

[22]Fried, pp. 475–493.

[23]De George, pp. 356–357.

Promulgation of "Standards for Privacy of Individually Identifiable Health Information: Final Rule"

James M. Humber

With the exception of this essay, all articles in this text were written early in 2000. At that time, the Department of Health and Human Services (HHS) was deep in the process of formulating federal standards for the protection of individually identifiable health information. Congress initiated this process in 1996 when it passed the Health Insurance Portability and Accountability Act (HIPAA). In this act, Congress required that steps be taken to improve "the efficiency and effectiveness of the health care system by encouraging the development of a health information system through the establishment of standards and requirements for the electronic transmission of certain health information."[1] In this mandate, Congress sought to achieve two goals: administrative simplification and cost reduction for electronically transmitted health information. At the same time, Congress recognized that these goals could not be achieved in the absence of a uniform set of standards for protecting the privacy and confidentiality of such medical records. Thus, in drafting the HIPAA, Congress required the Secretary of HHS to develop recommendations for medical privacy and to submit these recommendations to Congress for possible legislative action. It also specified that if Congress failed to enact legislation on this topic by August 21, 1999, the Secretary of HHS would be responsible for promulgating regulations containing standards with respect to the privacy of electronically transmitted, individually identifiable health information. Because Congress did not enact legislation by the assigned date, HHS published proposed privacy guidelines in the FEDERAL REGISTER on November 3, 1999 and called for comments on them.[2] Over 52,000 comments were received; after giving these comments due consideration, Donna E. Shalala, Secretary of HHS, promulgated final standards for protecting the privacy of Americans' personal health records.[3] These standards were published in the FEDERAL REGISTER on December 28, 2000, and were signed by executive order of President Clinton. The regulations take effect February 26, 2001. However, the entities to which the regulations apply have 24 months after that date to fully comply with the standards; thus, the final regulations do not come into full effect until February 26, 2003.[4]

The Final Standards for Privacy of Individually Identifiable Health Information as printed in the December 28, 2000 volume of the FEDERAL REGISTER are extremely lengthy. However, HHS has prepared a summary of these regulations which not only describes the principal features of the Final Rule for protecting the privacy of personal health records, but also explains how this Rule differs from the proposed standards that were published in the November 3, 1999 volume of the FEDERAL REGISTER.[5] This summary is as follows.

Protecting the Privacy of Patients' Health Information

Summary of the Final Regulation

Covered Entities

As required by HIPAA, the final regulation covers health plans, health care clearinghouses, and those health care providers who conduct certain financial and administrative transactions (e.g., electronic billing and funds transfers) electronically.

Information Protected

All medical records and other individually identifiable health information held or disclosed by a covered entity in any form, whether communicated electronically, on paper, or orally, is covered by the final regulation.

Components of the Final Rule

The rule is the result of the Department's careful consideration of every comment, and reflects a balance between accommodating practical uses of individually identifiable health information and rendering maximum privacy protection of that information.

Consumer Control Over Health Information

Under this final rule, patients have significant new rights to understand and control how their health information is used.

- Patient education on privacy protections: Providers and health plans are required to give patients a clear, written explanation of how they can use, keep, and disclose their health information.
- Ensuring patient access to their medical records: Patients must be able to see and get copies of their records, and request amendments. In addition, a history of most disclosures must be made accessible to patients. Receiving patient consent before information is released: Patient authorization to disclose information must meet specific requirements. Health care providers who see patients are required to obtain patient consent before sharing their information for treatment, payment, and health care operations purposes. In addition, specific patient consent must be sought and granted for non-routine uses and most non-health care purposes, such as releasing information to financial institutions determining mortgages and other loans or selling mailing lists to interested parties such as life insurers. Patients have the right to request restrictions on the uses and disclosures of their information.
- Ensuring that consent is not coerced: Providers and health plans generally cannot condition treatment on a

patient's agreement to disclose health information for non-routine uses.
- Providing recourse if privacy protections are violated: People have the right to complain to a covered provider or health plan, or to the Secretary, about violations of the provisions of this rule or the policies and procedures of the covered entity.

Boundaries on Medical Record Use and Release

With few exceptions, an individual's health information can be used for health purposes only.

- Ensuring that health information is not used for non-health purposes: Patient information can be used or disclosed by a health plan, provider, or clearinghouse only for purposes of health care treatment, payment, and operations. Health information is not used for non-health purposes: Patient information can be used or disclosed by a health plan, provider, or clearinghouse only for purposes of health care treatment, payment, and operations. Health information cannot be used for purposes not related to health care (such as use by employers to make personnel decisions, or use by financial institutions) without explicit authorization from the individual.
- Providing the minimum amount of information necessary: Disclosures of information must be limited to the minimum necessary for the purpose of the disclosure. However, this provision does not apply to the transfer of medical records for purposes of treatment, since physicians, specialists, and other providers need access to the full record to provide best quality care.
- Ensuring informed and voluntary consent: Non-routine disclosures with patient authorization must meet stan-

dards that ensure the authorization is truly informed and voluntary.

Ensure the Security of Personal Health Information

The regulation establishes the privacy safeguard standards that covered entities must meet, but it leaves detailed policies and procedures for meeting these standards to the discretion of each covered entity. In this way, implementation of the standards will be flexible and scalable, to account for the nature of each entity's business, and its size and resources. Covered entities must:

- Adopt written privacy procedures: These must include who has access to protected information, how it will be used within the entity, and when the information would or would not be disclosed to others. They must also take steps to ensure that their business associates protect the privacy of health information.
- Train employees and designate a privacy officer: Covered entities must provide sufficient training so that their employees understand the new privacy protections procedures, and designate an individual to be responsible for ensuring the procedures are followed.
- Establish grievance processes: Covered entities must provide a means for patients to make inquiries or complaints regarding the privacy of their records.

Establish Accountability for Medical Records Use and Release

Penalties for covered entities that misuse personal health information are provided in HIPAA.

- Civil penalties: Health plans, providers and clearinghouses that violate these standards would be subject to civil liability. Civil money penalties are $100 per incident, up to $25,000 per person, per year, per standard.

- Federal criminal penalties: There would be federal criminal penalties for health plans, providers, and clearinghouses that knowingly and improperly disclose information or obtain information under false pretenses. Penalties would be higher for actions designed to generate monetary gain. Criminal penalties are up to $50,000 and one year in prison for obtaining or disclosing protected health information; up to $100,000 and up to five years in prison for obtaining protected health information under "false pretenses;" and up to $250,000 and up to 10 years in prison for obtaining or disclosing protected health information with the intent to sell, transfer, or use it for commercial advantage, personal gain, or malicious harm.

Balancing Public Responsibility with Privacy Protections

After balancing privacy and other social values, HHS is establishing rules that would permit certain existing disclosures of health information without individual authorization for the following national priority activities and for activities that allow the health care system to operate more smoothly. All of these disclosures have been permitted under existing laws and regulations. Within certain guidelines found in the regulation, covered entities may disclose information for:

- Oversight of the health care system, including quality assurance activities
- Public health

- Research, generally limited to when a waiver of authorization is independently approved by a privacy board or Institutional Review Board
- Judicial and administrative proceedings
- Limited law enforcement activities
- Emergency circumstances
- For identification of the body of a deceased person, or the cause of death
- For facility patient directories
- For activities related to national defense and security

The rule permits, but does not require, these types of disclosures. If there is no other law requiring that information be disclosed, physicians and hospitals will still have to make judgments about whether to disclose information, in light of their own policies and ethical principles.

Special Protection for Psychotherapy Notes

Psychotherapy notes (used only by a psychotherapist) are held to a higher standard or protection because they are not part of the medical record and are never intended to be shared with anyone else. All other health information is considered to be sensitive and treated consistently under this rule.

Equivalent Treatment of Public and Private Sector Health Plans and Providers

The provisions of the final rule generally apply equally to private sector and public sector entities. For example, both private hospitals and government agency medical units must comply with the full range of requirements, such as providing notice, access rights, requiring consent before disclosure for routine uses, establishing contracts with business associates, among others.

Changes from the Proposed Regulation

- Providing coverage to personal medical records in all forms: The proposed regulation had applied only to electronic records and to any paper records that had at some point existed in electronic form. The final regulation extends protection to all types of personal health information created or held by covered entities, including oral communications and paper records that have not existed in electronic form. This creates a privacy system that covers virtually all health information held by hospitals, providers, health plans, and health insurers.

- Requiring consent for routine disclosures: The Final Rule requires most providers to obtain patient consent for routine disclosure of health records, in additional to requiring special patient authorization for non-routine disclosures. The earlier version had proposed allowing these routine disclosures without advance consent for purposes of treatment, payment, and health care operations (such as internal data gathering by a provider or health care plan). However, most individuals commenting on this provision, including many physicians, believed consent for these purposes should be obtained in advance, as is typically done today. The Final Rule retains the new requirement that patients must also be provided detailed written information on privacy rights and how their information will be used.

- Allowing disclosure of the full medical record to providers for purpose of treatment: For most disclosures, such as information submitted with bills, covered entities are required to send only the minimum information needed for the purpose of the disclosure. However, for purposes of treatment, providers need to be able to transmit fuller information. The Final Rule gives providers full discretion in determining what personal health information to include when sending patients' medical records to other providers for treatment purposes.

- Protecting against unauthorized use of medical records for employment purposes: Companies that sponsor health plans will not be able to access the personal health information held by the plan for employment-related purposes, without authorization from the patient.

Cost of Implementation

Recognizing the savings and cost potential of standardizing electronic claims processing and protecting privacy and security, Congress provided in HIPAA 1996 that the overall financial impact of the HIPAA regulations reduce costs. As such, the financial assessment of the privacy regulation includes the ten-year $29.9 billion savings HHS projects for the recently released electronic claims regulation and the projected $17.6 billion in costs projected for the privacy regulation. This produces a net savings of approximately $12.3 billion for the health care delivery system while improving the efficiency of health care, as well as privacy protection.

Preserving Existing, Strong State Confidentiality Laws

Stronger state laws (like those covering mental health, HIV infection, and AIDS information) continue to apply. These confidentiality protections are cumulative; the Final Rules sets a national "floor" of privacy standards that protect all Americans, but in some states, individuals enjoy additional protection. In circumstances where states have decided through law to require certain disclosures of health information for civic purposes, we do not preempt these mandates. The result is to give individuals the benefit of all laws providing confidentiality protection as well as to honor state priorities.

The Need for Further Congressional Action

HIPAA limits the application of our rule to the covered entities. It does not provide authority for the rule to reach many persons and businesses that work for covered entities or otherwise receive health information from them. So the rule cannot put in place appropriate restrictions on how such recipients of protected health information may use and redisclose such information. There is no statutory authority for a private right of action for individuals to enforce their privacy rights. We need Congressional action to fill these gaps in patient privacy protections.

Implementation of the Final Regulation

The final regulation will come into full effect in two years. The regulation will be enforced by HHS' Office for Civil Rights, which will provide assistance to providers, plans, and health clearinghouses in meeting the requirements of the regulation—including a toll free line to help answer questions: 1-866-OCR-PRIV (1-866-627-7748). The TTY number is 1-866-788-4989. A Website on the new regulation will also be available at http://www.hhs.gov/ocr.

As the above summary indicates, promulgation of the Final Rule for protecting health privacy information will not bring an end to discussion and debate on the issue of medical privacy. Even more important, however, are the strong indications that the Final Rule may not be truly final; for promulgation of the Rule has given rise to numerous calls for President Bush to rescind the regulations and consider the issue of health privacy anew.[†] These calls come from a variety of quarters. Some appear to be purely

[†]From the American Psychoanalytic Association, ". . . As previously reported, the Bush Administration has delayed the effective date of the health information privacy regulations from February 26, 2001 to April 14, 2001 and has provided for a comment period until March 30 on the final regulation." *See* p. 179 for update on HHS rules.

political and partisan in nature and without rational warrant. For example, Representatives Jack Kingston (R-Ga.), Tom Tancredo (R-Colo.), and George Radanovich (R-Calif.) have asked fellow members of Congress to sign a letter asking President Bush to rescind all the executive orders President Clinton signed after Election Day. Kingston supports this request by asserting:

> *We just feel like the legislative branch ought to be involved in these things like an equal branch of government....Some of these [executive orders] might be great ideas, but the process was circumvented....*[6]

However relevant Kingston's comments might be to some of the executive orders signed by President Clinton, it is clear they have no applicability to the order enacting the Final Rule for the Privacy of Individually Identifiable Health Information. After all, it was Congress (and a Republican controlled Congress at that), which directed the Secretary of HHS to promulgate health privacy standards in the event it failed to pass legislation by August 21, 1999, and which then failed to meet that deadline. Still, political pressure need not have a rational underpinning in order to be effective, and so it is possible that Kingston, Tancredo, and Radanovich will be successful in influencing the actions of the Bush administration.

Not all attacks upon the Final Rule are political in nature nor are they all clearly without merit. When comments on the proposed standards of November 3, 1999 were called for, some in the health care industry argued that although the HIPAA sought to achieve administrative simplification and a reduction in cost associated with medical claims, the proposed standards achieved neither of these goals[7]. Health care industry leaders have used the same arguments to attack the privacy standards embodied in the Final Rule. For example, although HHS estimates that implementation of the Final Rule will produce a net savings for the health care delivery system of approximately $12.3 billion over the next

10 years, the American Hospital Administration takes issue with this claim. It asserts that the HHS savings estimate is based on the assumption that it will cost the entire health care industry $17.6 billion over the next 10 years to comply with the regulations in the Final Rule, when in fact it will cost hospitals alone approximately $22.5 billion over the next 5 years to comply.[8] Moreover, if there is any merit at all to the claim that the proposed regulations of November 3, 1999 failed to enhance administrative simplification because these standards added "complexities to ensuring the free flow of information among certain healthcare professionals and institutions trying to coordinate patient care,"[7] this argument has even greater force when directed against the Final Rule. This must be acknowledged because the final set of regulations extends privacy protection beyond the electronic and paper records covered by the proposed regulations and mandates privacy protection for all types of personal health information created or held by covered entities. Furthermore, the Final Rule requires providers of health care to obtain patient consent for routine disclosures of health records, whereas the proposed regulations did not. Whether or not these considerations will cause the Bush administration to rescind the Final Rule cannot be known at this time. However, they appear to have had some effect; for Ari Fleischer, a Bush spokesman, says that the Bush administration will at least review the Final Regulations.[9]

Aside from the above, the Final Standards for the Protection of Individually Identifiable Health Information seem open to attack in at least two other ways. First, the standards allow covered entities to be liable for civil penalties whenever they violate provisions of the Final Rule. At the same time, they tell covered entities that "disclosures of information must be limited to the minimum necessary for the purposes of disclosure" and that "the regulation establishes privacy safeguard standards that covered entities must meet, but it leaves detailed policies and procedures for meeting these standards to the discretion of each covered entity." These "guidelines" appear to provide no real guidance at all for covered entities seeking to put their practices in accord with the privacy safeguard standards of the Final Rule. How, after all, are covered entities to determine the "minimum necessary" information for each individual transfer of medical records? And how are they to know whether or not their policies and procedures for ensuring privacy are sufficient for meeting the standards of the Final Rule? Indeed, if HHS had wanted to craft a document specifically designed to throw open the proverbial and much maligned "floodgates of litigation," it is difficult to see how they could have done a better job.

Second, the Final Rule also seems vulnerable in its imposition of criminal penalties. Hackers already have gained access to some hospital records, and they have even broken into systems of computer corporations such as Microsoft and Yahoo. Also, unscrupulous and dishonest people exist in every corporation, every health care enterprise, and every government agency. Thus, there seems little doubt that electronically stored, individually identifiable health information is vulnerable to access by unauthorized and/or criminal sources. Moreover, the financial rewards for gaining access to this information could be tremendous. Given these circumstances one can doubt that the criminal penalties imposed by the Final Rule upon those who improperly obtain or disclose protected medical information are severe enough to discourage theft of medical records. If these penalties are not sufficient to deter at least a significant number of assaults upon medical privacy, all of the efforts and resources expended in the effort to promulgate, implement, and adjudicate the health privacy standards will have been for naught. Imposing much more severe criminal penalties for violations of medical privacy would cost society very little, and the gains could be great. This being so, it is difficult to se why the criminal sanctions assessed by the Final Rule for the Protection of Individually Identifiable Health Information should not be strengthened.

Notes and References

[1]"Standards for Privacy of Individually Identifiable Health Information: Final Rule," 65 CFR, Sec. 160.101, Dec. 28, 2000, p. 82463.

[2]"Standards for Privacy of Individually Identifiable Health Information," 45 CFR, Parts 160 and 164, Nov. 3, 1999.

[3]65 CFR, op. cit., pp. 82461–82510.

[4]There is one exception to this assertion. Small health plans have 36 months to comply with the Final Regulations (*see ibid.,* p. 82470). Thus, for these covered entities, the Final Rule does not become fully operative until 2/26/04.

[5]*See* "HHS Fact Sheet," http://aspe.hhs.gov/admnsimp/pvcfact1.htm, Jan. 17, 2001.

[6]"GOP Trio Targets Clinton Orders," *Atlanta Constitution,* Jan. 20, 2001, p. A5.

[7]DeMuro, Paul and Gantt III, W. Andrew, "HIPAA Privacy Standards Raise Complex Implementation Issues," *Health Care Financial Management,* January, 2001, pp. 42–47.

[8]Jaspen, Bruce, "Records Law May Sicken Care Firms," *Chicago Tribune,* Dec. 29, 2000, Bus. Sec., p. 1.

[9]Crowley, Susan, "New Privacy Rules Touch Off Dispute," *AARP Bulletin,* February, 2001, p. 3.

[†]Current status at presstime is that HHS Secretary Thompson issued a new 30-day public comment period on February 23, 2001, thus delaying implementation of the final rules until mid-April. *See* following *Washington Post* article for a reference (Margo Goldman):

Patient Privacy Rules Put on Hold

Saturday, February 24, 2001; Page A09

Health and Human Services Secretary Tommy G. Thompson announced yesterday that he would allow a new round of outside critiques of the first federal standards to safeguard the confidentiality of patients' medical records.

The regulations were to have taken effect next week, after years of controversy among consumer activists and health care interests. Thompson's creation of the unexpected, 30-day comment period gave hope to longstanding opponents of the rules, including the insurance industry and the American Medical Association.

In a statement, Thompson said the privacy rules could not take effect, as planned, even if he had not agreed to reconsider them. He said that was because Clinton administration officials inadvertently did not send a copy of the rules to Congress two months ago, when they were published in the Federal Register. The rules thus cannot be implemented until mid-April.

The secretary gave no hint if he plans to change the regulations after evaluating the new critiques. HHS was required by law to develop the standards after Congress missed a self-imposed deadline for adopting such rules.

—Amy Goldstein

© 2001 The Washington Post Company

Index